CLIMATE POLICY

International Perspectives on Greenhouse Gases

CLIMATE POLICY
International Perspectives on Greenhouse Gases

Edited by
Gabriela Ionescu, PhD

Apple Academic Press Inc.
3333 Mistwell Crescent
Oakville, ON L6L 0A2
Canada

Apple Academic Press Inc.
9 Spinnaker Way
Waretown, NJ 08758
USA

©2016 by Apple Academic Press, Inc.

First issued in paperback 2021

Exclusive worldwide distribution by CRC Press, a member of Taylor & Francis Group

No claim to original U.S. Government works

ISBN 13: 978-1-77463-699-2 (pbk)
ISBN 13: 978-1-77188-414-3 (hbk)

Library and Archives Canada Cataloguing in Publication

Climate policy : international perspectives on greenhouse gases / edited by Gabriela Ionescu, PhD.

Includes bibliographical references and index.
Issued in print and electronic formats.
ISBN 978-1-77188-414-3 (bound).--ISBN 978-1-77188-415-0 (pdf)
1. Greenhouse gas mitigation. 2. Climate change mitigation.
3. Climatic changes. 4. Global warming. 5. Renewable energy sources.
I. Ionescu, Gabriela, editor

QC903.C567 2016   363.738'74   C2016-900851-7   C2016-900852-5

Library of Congress Cataloging-in-Publication Data

Names: Ionescu, Gabriela, editor.
Title: Climate policy : international perspectives on greenhouse gases/Gabriela Ionescu, PhD.
Description: Toronto : Apple Academic Press, 2016. | Includes bibliographical references and index.
Identifiers: LCCN 2016004877 (print) | LCCN 2016011675 (ebook) | ISBN 9781771884143 (hardcover : alk. paper) | ISBN 9781771884150 ()
Subjects: LCSH: Atmospheric carbon dioxide--Environmental aspects. | Greenhouse gases--Environmental aspects. | Air--Pollution--Environmental aspects. | Climatic changes. | Sustainable development.
Classification: LCC QC879.8 .C585 2016 (print) | LCC QC879.8 (ebook) | DDC 363.738/74561--dc23
LC record available at http://lccn.loc.gov/2016004877

Apple Academic Press also publishes its books in a variety of electronic formats. Some content that appears in print may not be available in electronic format. For information about Apple Academic Press products, visit our website at **www.appleacademicpress.com** and the CRC Press website at **www. crcpress.com**

ABOUT THE EDITOR

GABRIELA IONESCU, PhD

Dr. Gabriela Ionescu obtained her PhD in Power Engineering from Politehnica University of Bucharest and in Environmental Engineering from University of Trento. She is currently a member of the Department of Energy Production and Use at the Politehnica University of Bucharest and collaborator of the Department of Civil, Environmental and Mechanical Engineering at the University of Trento. She has done prolific research and has been published multiple times in areas related to energy efficiency, waste and wastewater management, energy conservation, life-cycle assessment, environmental analysis, and sustainability feasibility studies.

CONTENTS

Part IV: Fossil Fuel Alternatives

Part V: Conclusions

ACKNOWLEDGMENT AND HOW TO CITE

The editor and publisher thank each of the authors who contributed to this book. The chapters in this book were previously published elsewhere. To cite the work contained in this book and to view the individual permissions, please refer to the citation at the beginning of each chapter. Each chapter was read individually and carefully selected by the editor; the result is a book that provides a multiperspective look at ways to reduce greenhouse gases and find fossil fuel alternatives. The chapters included examine the following topics:

Part I: Climate Change Impact
- Chapter 1 proposes a new metric for comparing greenhouse gases, termed climate change impact potential, as an alternative to global warming potential.

Part II: Reducing Carbon Footprint
- Chapter 2 is based on the systematization of information on the directives of the European Commission concerning the reduction in the use of natural resources, the processing of sets of relevant statistical indicators, and the analysis and identification of solutions suitable to mitigate the impact of these carbon dioxide emissions through the use of renewable resources.
- Chapter 3 is a study on Arun 160 technology, the only paraboloid concentrator technology developed in India till date for industrial process heat applications, and the barriers it faces in its market deployment despite the economic and environmental advantages it offers.

Part III: Other Greenhouse Gases

- Chapter 4 reviews studies of climate-chemistry interactions affecting current and future distributions of atmospheric ozone and methane.
- Chapter 5 builds upon the RX10 study to further account for the newly developed projections of HFC emissions and provides a detailed analysis of the implication of HFC mitigation on global temperature.
- Chapter 6 compares the thermodynamic performances of the direct expansion ground source heat pump system using three kinds of refrigerants and recommends a suitable alternative substance for HFC-134a based on theoretical calculation and analysis.
- Chapter 7 integrates the contributions of the authors to the UNEP report, other chapters in that report, the Intergovernmental Panel on Climate Change Report fifth assessment (AR5), and other recent literature for the most recent information on N_2O.

Part IV: Fossil Fuel Alternatives

- Chapter 8 sketches the issues of allocation and how it has been dealt with in policy guidelines on bioenergy, describing a case study on electricity with rapeseed that uses several allocation principles.
- Chapter 9 proposes an improved optimal design method for stand-alone wind-solar-battery hybrid power system.
- Chapter 10 adds to the existing literature by addressing not only the effect of global temperature potential, but also the immediate policy-relevant question of using 100-year global warming potential values from various assessment reports and global warming potentials calculated over different time spans

Part V: Conclusions

- Chapter 11 asks the question how the formulation of climate policy scenarios could be framed to enhance their applicability to the scenario framework, in particular concerning their introduction into the representative concentration pathways–socioeconomic reference pathways matrix.

LIST OF CONTRIBUTORS

Pleasa Serin Abraham
Department of Humanities & Social Sciences, IIT Bombay, Mumbai, India

Angelica Mendoza Beltran
Netherlands Environmental Assessment Agency, P.O. Box 303, Bilthoven, The Netherlands

Maarten van den Berg
PBL Netherlands Environmental Assessment Agency, Bilthoven, the Netherlands

Terje K. Berntsen
Department of Geosciences, University of Oslo, N-0315 Oslo, Norway; Center for International Climate and Environmental Research—Oslo (CICERO), N-0318 Oslo, Norway

Lucian-Ionel Cioca
Department of Industrial Engineering and Management, Faculty of Engineering, "Lucian Blaga" University of Sibiu, Bd. Victoriei No.10, 550024 Sibiu, Romania

Stig B. Dalsøren
Center for International Climate and Environmental Research—Oslo (CICERO), N-0318 Oslo, Norway

Eric A. Davidson
The Woods Hole Research Center, 149 Woods Hole Road, Falmouth, MA 02540-1644, USA

Kristie L. Ebi
ClimAdapt,LLC, Los Altos, CA, USA

Jae Edmonds
Joint Global Change Research Institute, Pacific Northwest National Laboratory, College Park, MD, USA

Kostas Eleftheratos
Department of Geology and Geoenvironment, University of Athens, 15784 Athens, Greece

Yuefen Gao
School of Energy, Power & Mechanical Engineering, North China Electric Power University, Baoding, China; Swan Center for Energy Research, Newcastle University, Newcastle Upon Tyne, UK

Jeroen Guinée
Institute of Environmental Sciences, Leiden University, P.O. Box 9518, Leiden, The Netherlands

Haripriya Gundimeda
Department of Humanities & Social Sciences, IIT Bombay, Mumbai, India

Stéphane Hallegatte
The World Bank, Washington, DC, USA

Reinout Heijungs
Institute of Environmental Sciences, Leiden University, P.O. Box 9518, Leiden, The Netherlands

Andries F. Hof
PBL Netherlands Environmental Assessment Agency, Bilthoven, the Netherlands

Chris D. Holmes
University of California, Irvine, CA 92697, USA

Gabriela Ionescu
Department of Energy Production and Use, Politehnica University of Bucharest, Splaiul Independentei 313, Sector 6, 060042 Bucharest, Romania

Ivar S. A. Isaksen
Department of Geosciences, University of Oslo, N-0315 Oslo, Norway; Center for International Climate and Environmental Research—Oslo (CICERO), N-0318 Oslo, Norway

Larisa Ivascu
Department of Management, Faculty of Management in Production and Transportation, Politehnica University of Timisoara, Piata Victoria No.2, 300006 Timisoara, Romania

Yonggang Jiao
School of Electrical Engineering, Shenyang University of Technology, Shenyang 110870, China

Zhenhe Ju
School of Renewable Energy, Shenyang Institute of Engineering, Shenyang 110136, China

David Kanter
The Earth Institute, Columbia University, 535 West 116th Street, New York, NY 10027, USA

Miko U. F. Kirschbaum
Landcare Research, Private Bag 11052, Palmerston North 4442, New Zealand

Tom Kram
PBL Netherlands Environmental Assessment Agency, The Hague, The Netherlands

Elmar Kriegler
Potsdam Institute of Climate Impact Research, Potsdam, Germany

Yvan Orsolini
NORWEGIAN Institute for Air Research (NILU), N-2027 Kjeller, Norway

Yingxin Peng
School of Energy, Power & Mechanical Engineering, North China Electric Power University, Baoding, China

Elena Cristina Rada
Department of Biotechnologies and Life Sciences, Insubria University of Varese, via G.B. Vico 46, I-21100 Varese, Italy; Department of Civil Environmental and Mechanical Engineering, Italy & University of Trento, via Mesiano 77, 38123 Trento, Italy

V. Ramanathan
Scripps Institution of Oceanography, UC San Diego, 9500 Gilman Dr., La Jolla, CA 92093, USA

Keywan Riahi
International Institute for Applied System Analysis, Laxenburg, Austria

Bjørg Rognerud
Department of Geosciences, University of Oslo, N-0315 Oslo, Norway

Tony Roskilly
Swan Center for Energy Research, Newcastle University, Newcastle Upon Tyne, UK

Theo van Ruijven
Faculty of Technology, Policy and Management, Delft University of Technology, P.O. Box 5015, Delft, The Netherlands

Ole Amund Søvde
Center for International Climate and Environmental Research—Oslo (CICERO), N-0318 Oslo, Norway

Frode Stordal
Department of Geosciences, University of Oslo, N-0315 Oslo, Norway

Vincenzo Torretta
Department of Biotechnologies and Life Sciences, Insubria University of Varese, via G.B. Vico 46, I-21100 Varese, Italy

Kathrine Vad
Institute of Environmental Sciences, Leiden University, P.O. Box 9518, Leiden, The Netherlands

G. J. M. Velders
National Institute for Public Health and the Environment (RIVM), P.O. Box 1, 3720 BA Bilthoven, the Netherlands

Jasper van Vliet
PBL Netherlands Environmental Assessment Agency, Bilthoven, the Netherlands

Detlef P. van Vuuren
PBL Netherlands Environmental Assessment Agency, Bilthoven, the Netherlands; Utrecht University, Department of Geosciences, Copernicus Institute of Sustainable Development, Utrecht, the Netherlands

Tjerk Wardenaar
Science System Assessment, Rathenau Instituut, P.O. Box 95366, The Hague, The Netherland

Harald Winkler
Energy Research Centre, University of Cape Town, Cape Town, Republic of South Africa

Y. Xu
Scripps Institution of Oceanography, UC San Diego, 9500 Gilman Dr., La Jolla, CA 92093, USA

Xiaoju Yin
School of Electrical Engineering, Shenyang University of Technology, Shenyang 110870, China;
School of Renewable Energy, Shenyang Institute of Engineering, Shenyang 110136, China

D. Zaelke
Program on Governance for Sustainable Development, Bren School of Environmental Science &
Management, UC Santa Barbara, CA 93106, USA

Christos Zerefos
Academy of Athens, 10680 Athens, Greece

Fengge Zhang
School of Electrical Engineering, Shenyang University of Technology, Shenyang 110870, China

Honglei Zhao
School of Energy, Power & Mechanical Engineering, North China Electric Power University, Baoding,
China

INTRODUCTION

Climate change is a reality all over the world, and its complexity is increasing. Therefore, sustainability has become a national and international concern, ingrained in many organizational processes. The ability of organizations to respond to sustainability concerns is sometimes hindered by the complexity of integrating sustainability into business models and by the need to rethink their strategic directions.

For policy applications, such as for the Kyoto Protocol, the climate-change contributions of different greenhouse gases are usually quantified through their global warming potentials. They are calculated based on the cumulative radiative forcing resulting from a pulse emission of a gas over a specified time period. However, these calculations are not explicitly linked to an assessment of ultimate climate-change impacts. A new metric, the climate-change impact potential (CCIP), is presented in chapter 1 that is based on explicitly defining the climate-change perturbations that lead to three different kinds of climate-change impacts. These kinds of impacts are: (1) those related directly to temperature increases; (2) those related to the rate of warming; and (3) those related to cumulative warming. From those definitions, a quantitative assessment of the importance of pulse emissions of each gas is developed, with each kind of impact assigned equal weight for an overall impact assessment. Total impacts are calculated under the RCP6 concentration pathway as a base case. The relevant climate-change impact potentials are then calculated as the marginal increase of those impacts over 100 years through the emission of an additional unit of each gas in 2010. These calculations are demonstrated for carbon dioxide, methane, and nitrous oxide. Compared with global warming potentials, climate-change impact potentials would increase the importance of pulse emissions of long-lived nitrous oxide and reduce the importance of short-lived methane.

In Romania, sustainable development has become a priority for businesses, but even though companies are showing some concern, there are yet to demonstrate any full commitment (they are mainly concerned with areas such as society and the environment). Chapter 2 assesses Romania's involvement in the adoption of actions directed toward the reduction of pollutants and greenhouse gases, namely actions focused on reducing the main causes of pollution. This analysis compares the situation in Romania with that of the European Union. The main concerns can be categorized according to four sectors, which produce the highest quantity of carbon dioxide emissions in the world: the energy sector, the transport sector, the waste sector and the industry sector. The last section of this chapter deals with the carbon footprint of Romania and its implications.

The persistent market failures and policy inertia due to the existence of carbon lock-in create barriers to the diffusion of carbon saving technologies. In spite of their apparent environmental and technological advantages, the renewable technologies cannot take off in the market. Chapter 3 studies on an empirical level the barriers to the market deployment of indigenously developed Arun™ 160 Solar Concentrator technology, which has the potential to revolutionize the industrial restructuring by providing solar heat for industrial processes. The authors classify the barriers into micro, meso, and macro barriers, and analyze the impact of them by conducting two rounds of questionnaires: one with firms and other with experts.

Ozone and methane are chemically active climate-forcing agents affected by climate–chemistry interactions in the atmosphere. Key chemical reactions and processes affecting ozone and methane are presented in chapter 4. The authors show that climate-chemistry interactions have a significant impact on the two compounds. Ozone, which is a secondary compound in the atmosphere, produced and broken down mainly in the troposphere and stratosphere through chemical reactions involving atomic oxygen (O), NOx compounds (NO, NO_2), CO, hydrogen radicals (OH, HO_2), volatile organic compounds (VOC) and chlorine (Cl, ClO), and bromine (Br, BrO). Ozone is broken down through changes in the atmospheric distribution of these compounds. Methane is a primary compound emitted from different sources (wetlands, rice production, livestock, mining, oil and gas production, and landfills). Methane is broken down by the hydroxyl radical (OH). OH is significantly affected by methane emissions, defined by the feedback factor, currently estimated to be in the range 1.3 to 1.5, and increasing with increasing methane emission. Ozone and methane changes are affected by NOx emissions. While ozone in general increase with increases in NOx emission, methane is reduced, due to increases in OH. The authors identify several processes where current and future changes have implications for climate-chemistry interactions. They also show that climatic changes through dynamic processes could have significant impact on the atmospheric chemical distribution of ozone and methane, as can be seen through the impact of Quasi Biennial Oscillation (QBO). Modeling studies indicate that increases in ozone could be more pronounced toward the end of this century. Thawing permafrost could lead to important positive feedbacks in the climate system. Large amounts of organic material are stored in the upper layers of the permafrost in the yedoma deposits in Siberia, where 2 to 5% of the deposits could be organic material. During thawing of permafrost, parts of the organic material that is deposited could be converted to methane. Furthermore, methane stored in deposits under shallow waters in the Arctic have the potential to be released in a future warmer climate with enhanced climate impact on methane, ozone, and stratospheric water vapor. Studies performed by several groups show that the transport sectors have the potential for significant impacts on climate-chemistry interactions. There are large uncertainties connected to ozone and methane changes from the transport sector, and to methane release and climate impact during permafrost thawing.

There is growing international interest in mitigating climate change during the early part of this century by reducing emissions of short-lived climate pollutants (SLCPs), in addition to reducing emissions of carbon dioxide. The SLCPs include methane (CH_4), black carbon aerosols (BC), tropospheric ozone (O_3) and hydrofluorocarbons (HFCs). Recent studies have estimated that by mitigating emissions of CH4, BC, and O_3 using available technologies, about 0.5 to 0.6 °C warming can be avoided by mid-21st century. In chapter 5, the authors show that avoiding production and use of high-GWP (global warming potential) HFCs by using technologically feasible low-GWP substitutes to meet the increasing global demand can avoid as much as another 0.5 °C warming by the end of the century. This combined mitigation of SLCPs would cut the cumulative warming since 2005 by 50% at 2050 and by 60% at 2100 from the CO_2-only mitigation scenarios, significantly reducing the rate of warming and lowering the probability of exceeding the 2°C warming threshold during this century.

HFO-1234yf and HFO-1234ze[E] have low global warming potential and zero ozone depletion potential. If they are used in the direct expansion ground source heat pump system substituting for HFC-134a, the system will be beneficial to mitigating climate change. Chapter 6 aims to find out the thermodynamic characteristics of the direct expansion ground source heat pump system using HFO-1234yf or HFO-1234ze[E] by theoretical calculation. The results indicate that HFO-1234yf system in an actual cycle has the highest COP. HFO-1234yf and HFO-1234ze[E] have such smaller capacity per unit of swept volume that they need larger compression capacity if providing the same heating or cooling loads. For a given unit when HFC-134a is replaced with HFO-1234yf or HFO-1234ze[E], the capacity will decrease. More refrigerant charge is required in the HFO-1234yf or HFO-1234ze[E] system. The results also present that more refrigerant charge is required in the cooling mode than in the heating mode.

Effective mitigation for N_2O emissions, now the third most important anthropogenic greenhouse gas and the largest remaining anthropogenic source of stratospheric ozone depleting substances, requires understanding the sources and how they may increase this century. In chapter 7, the authors update estimates and their uncertainties for current anthropogenic and natural N_2O emissions and for emissions scenarios to 2050. Although major uncertainties remain, "bottom-up" inventories and "top-down" atmospheric modeling yield estimates that are in broad agreement. Global natural N_2O emissions are most likely between 10 and 12 Tg N_2O-N yr^{-1}. Net anthropogenic N_2O emissions are now about 5.3 Tg N_2O-N yr^{-1}. Gross anthropogenic emissions by sector are 66% from agriculture, 15% from energy and transport sectors, 11% from biomass burning, and 8% from other sources. A decrease in natural emissions from tropical soils due to deforestation reduces gross anthropogenic emissions by about 14%. Business-as-usual emission scenarios project almost a doubling of anthropogenic N_2O emissions by 2050. In contrast, concerted mitigation scenarios project an average decline of 22% relative to 2005, which would lead to a near stabilization of atmospheric concentration of N_2O at about 350 ppb. The impact of growing demand for biofuels on future projections of N_2O emissions is highly uncertain; N_2O emissions from second and third generation biofuels could remain trivial or could become the most significant source to date. It will not be possible to completely eliminate anthropogenic N_2O emissions from agriculture, but better matching of crop N needs and N supply offers significant opportunities for emission reductions.

The increasing concern for adverse effects of climate change has spurred the search for alternatives for conventional energy sources. Life cycle assessment (LCA) has increasingly been used to assess the potential of these alternatives to reduce greenhouse gas emissions. The popularity of LCA in the policy context puts its methodological issues into another perspective. Chapter 8 discusses how bio-electricity directives deal with the issue of allocation and shows its repercussions in the policy field. Multifunctionality has been a well-known problem since the early development of LCA and several methods have been suggested to deal with multifunctional processes. The chapter starts with a discussion of the most common allocation methods. This discussion is followed by a description of bio-energy policy directives. The description shows the increasing importance of LCA in the policy context as well as the lack of consensus in the application of allocation methods. Methodological differences between bio-energy directives possibly lead to different assessments of bio-energy chains. To assess the differences due to methodological choices in bio-energy directives, the authors apply three different allocation methods to the same bio-electricity generation system. The differences in outcomes indicate the importance of solving the allocation issue for policy decision making. The case study focuses on bio-electricity from rapeseed oil. To assess the influence of the choice of allocation in a policy directive, three allocation methods are applied: economic partitioning (on the basis of proceeds), physical partitioning (on the basis of energy content), and substitution (under two scenarios). The outcomes show that the climate change score is assessed quite differently, ranging from 0.293 kg to 0.604 kg CO_2 eq/kWh. The authors argue that this uncertainty hampers the optimal use of LCA in the policy context. The aim of policy LCAs is different from the aim of LCAs for analysis. Therefore, the authors argue that LCAs in the policy context will benefit from a new guideline based on robustness. The case study confirms that the choice of allocation method in policy directives has great influence on the outcomes of an LCA. With the growing popularity of LCA in policy directives, this chapter recommends a new guideline for policy LCAs. The high priority of robustness in the policy context makes it an ideal starting point of this guideline. An accompanying dialog between practitioners and commissioners should further strengthen the use of LCA in policy directives.

The reliability and economic value of wind and solar power generation system with energy storage are decided by the balance of capacity distribution. The improved capacity balance matching method is proposed in chapter 9, which not only utilizes the complementary characteristics of the wind and solar power generation system sufficiently but also reduces the charge and discharge times of the battery. Therefore, the generation reliability is improved and the working lifetime of the whole system is lengthened. Consequently, the investment of the battery energy storage is reduced as well as the whole cost is decreased. The experimental result was presented to verify the effectiveness of the improved optimal capacity ratio design method.

Chapter 10 analyzes the effect of different emission metrics and metric values on timing and costs of greenhouse gas mitigation in least-cost emission pathways aimed at a forcing level of 3.5 W m^{-2} in 2100. Such an assessment is currently relevant in view of UNFCCC's decision to replace the values currently used. An emission metric determines the relative weights of non-CO_2 greenhouse gases in obtaining CO_2-equivalent emissions. For the first commitment period of the Kyoto Protocol, the UNFCCC has used 100-year GWP values as reported in IPCC's Second Assessment Report. For the second commitment period, the UNFCCC has decided to use 100 year GWP values from IPCC's Fourth Assessment Report. The authors find that such a change has only a minor impact on (the optimal timing of) global emission reductions and costs. However, using 20 year or 500 year GWPs to value non-CO_2 greenhouse gases does result in a significant change in both costs and emission reductions in our model. CO_2 reductions are favored over non-CO_2 gases when the time horizon of the GWPs is increased. Application of GWPs with time horizons longer than 100 years can increase abatement costs substantially, by about 20% for 500 year GWPs. Surprisingly, the authors find that implementation of a metric based on a time-dependent global temperature potential does not necessary lead to lower abatement costs. The crucial factor here is how fast non-CO_2 emissions can be reduced; if this is limited, the delay in reducing methane emissions cannot be (fully) compensated for later in the century, which increases total abatement costs.

The new scenario framework facilitates the coupling of multiple socioeconomic reference pathways with climate model products using the representative concentration pathways. This will allow for improved assessment of climate impacts, adaptation, and mitigation. Assumptions about climate policy play a major role in linking socioeconomic futures with forcing and climate outcomes. Chapter 11 presents the concept of shared climate policy assumptions as an important element of the new scenario framework. Shared climate policy assumptions capture key policy attributes such as the goals, instruments, and obstacles of mitigation and adaptation measures, and introduce an important additional dimension to the scenario matrix architecture. They can be used to improve the comparability of scenarios in the scenario matrix. Shared climate policy assumptions should be designed to be policy relevant, and as a set to be broad enough to allow a comprehensive exploration of the climate change scenario space.

Climate change *will* happen; now we must find ways to slow down the process. Climate change mitigation will require an international multifaceted approach to move the world toward a low-carbon society that produces less overall greenhouse gas. Given the immensity of the crisis the world faces, research must be ongoing.

PART I

CLIMATE CHANGE IMPACT

CHAPTER 1

Climate-Change Impact Potentials as an Alternative to Global Warming Potentials

MIKO U. F. KIRSCHBAUM

1.1 INTRODUCTION

Climate-change policies aim to prevent ultimate adverse climate-change impacts, stated explicitly by the UNFCCC as 'preventing dangerous anthropogenic interference with the climate system'. This has led to the adoption of specific climate-change targets to avoid exceeding certain temperature thresholds, such as the '2° target' agreed to in Copenhagen in 2009. The UNFCCC also stated that this aim should be achieved through measures that are 'comprehensive and cost-effective'. To achieve comprehensive and cost-effective climate-change mitigation requires an assessment of the relative marginal contribution of different greenhouse gases (GHGs) to ultimate climate-change impacts.

Currently, the importance of the emission of different GHGs is usually quantified through their global warming potentials (GWPs), which are calculated as their cumulative radiative forcing over a specified time horizons under constant GHG concentrations (e.g. Lashof and Ahuja 1990, Fuglestvedt et al 2003). Typical time horizons are 20, 100 and 500 years, with 100 years used most commonly, such as for the Kyoto Protocol. Setting targets in terms of avoiding specified peak temperatures is, however, conceptually inconsistent with a metric that is based on cumulative radiative forcing (e.g. Smith et al 2012). Climate-change metrics were also discussed at a 2009 IPCC expert workshop that noted shortcomings of GWPs and laid out requirements for appropriate metrics, but proposed no alternatives (Plattner et al 2009). Other important issues related to GHG accounting were discussed by Manne and Richels (2001), Fuglestvedt et al (2003, 2010), Johansson et al (2006), Tanaka et al (2010), Peters et al (2011a, 2011b), Manning and Reisinger (2011), Johansson (2012), Kendall (2012), Ekholm et al (2013) and Brandão et al (2013).

Out of these and earlier discussions emerged proposals for alternative metrics. Most prominent among these is the global temperature change potential (GTP), proposed by Shine et al (2005, 2007), which is based on assessing the temperature that might be reached in future years and can be linked directly to adopted temperature targets. A key difference between GWPs and GTPs is that GWPs are measures of the cumulative GHG impact, whereas GTPs are measures of the direct or instantaneous GHG impact. Some impacts, most notably sea-level rise, are not functions of the temperature in future years, but of the cumulative warming leading up to those years (Vermeer and Rahmstorf 2009). Even if the global temperature were to reach and then stabilized at 2 °C above pre-industrial levels, sea levels would continue to rise for centuries (Vermeer and Rahmstorf 2009, Meehl et al 2012). Mitigation efforts that focus solely on maximum temperature increases thus provide no limit on future sea levels rise and only partly address the totality of climate-change impacts.

To be consistent with the policy aim of preventing adverse climate-change impacts, GHG metrics must include all relevant impacts. It is therefore necessary to explicitly define the climate-change perturbations that lead to specific kinds of impacts. The present paper proposes a new metric for comparing GHGs as an alternative to GWPs, termed climate-change impact potential (CCIP). It is based on an explicit definition and quantification of the climate perturbations that lead to different kinds of climatic impacts.

1.2 REQUIREMENTS OF AN IMPROVED METRIC

1.2.1 KINDS OF CLIMATE-CHANGE IMPACTS

There are at least three different kinds of climate-change impacts (Kirschbaum 2003a, 2003b, 2006, Fuglestvedt et al 2003, Tanaka et al 2010) that can be categorized based on their functional relationship to increasing temperature as:

(1) the impact related directly to elevated temperature;
(2) the impact related to the rate of warming; and
(3) the impact related to cumulative warming.

1.2.1.1 DIRECT-TEMPERATURE IMPACTS

Impacts related directly to temperature increases are easiest to focus on, and are the basis of the notion of keeping warming to 2 °C above pre-industrial temperatures. It is also the explicit metric for calculating GTPs (Shine et al 2005). It is the relevant measure for impacts such as heat waves (e.g. Huang et al 2011) and other extreme weather events (e.g. Webster et al 2005). Coral bleaching, for example, has occurred in nearly all tropical coral-growing regions and is unambiguously related to increased temperatures (e.g. Baker et al 2008).

1.2.1.2 RATE-OF-WARMING IMPACTS

The rate of warming is a concern because higher temperatures may not be inherently worse than cooler conditions, but change itself will cause problems for both natural and socio-economic systems. A slow rate of change will allow time for migration or other adjustments, but faster rates of change may give insufficient time for such adjustments (e.g. Peck and Teisberg 1994).

For example, the natural distribution of most species is restricted to narrow temperature ranges (e.g. Hughes et al 1996). As climate change makes their current habitats climatically unsuitable for many species (Parmesan and Yohe 2003), it poses serious and massive extinction risks (e.g. Thomas et al 2004). The rate of warming will strongly influence whether species can migrate to newly suitable habitats, or whether they will be driven to extinction in their old habitats.

1.2.1.3 CUMULATIVE-WARMING IMPACTS

The third kind of impact includes impacts such as sea-level rise (Vermeer and Rahmstorf 2009) which is quantified by cumulative warming, as sea-level rise is related to both the magnitude of warming and the length of time over which oceans and glaciers are exposed to increased temperatures. Lenton et al (2008) listed some possible tipping points in the global climate system, including shut-off of the Atlantic thermohaline circulation and Arctic sea-ice melting. If the world passes these thresholds, the global climate could shift into a different mode, with possibly serious and irreversible consequences. Their likely occurrence is often linked to cumulative warming. Cumulative warming is similar to the calculation of GWPs except that GWPs integrate only radiative forcing without considering the time lag between radiative forcing and resultant effects on global temperatures. The difference between GWPs and integrated warming are, however, only small over a 100-year time horizon and diminish even further over longer time horizons (Peters et al 2011a).

1.2.2 THE RELATIVE IMPORTANCE OF DIFFERENT KINDS OF IMPACTS

For devising optimal climate-change mitigation strategies, it is also necessary to quantify the importance of different kinds of impacts relative to each other. Without any formal assessment of their relative importance being available in the literature, they were therefore assigned here the same relative weighting. However, the different kinds of impacts change differently over time so that the importance of one kind of impact also changes over time relative to the importance of the others.

The notion of assigning them equal importance can therefore be implemented mathematically only under a specified emission pathway and at a defined point in time. This was done by expressing each impact relative to the most severe impact over the next 100 years under the 'representative concentration pathway' (RCP) with radiative forcing of 6 W m^{-2} (RCP6; van Vuuren et al 2011).

1.2.3 CUMULATIVE DAMAGES OR MOST SEVERE DAMAGES?

Any focus on maximum temperature increases, such as the '2° target', explicitly targets the most extreme impacts. However, that ignores the lesser, but still important, impacts that occur before and after the most extreme impacts are experienced. Hence, the damage function used here sums all impacts over the next 100 years. Summing impacts is different from summing temperatures to derive initial impacts. For example, the damage from tropical cyclones is linked to sea-surface temperatures in a given year (Webster et al 2005). Total damages to society, however, are the sum of cyclone damages in all years over the defined assessment horizon.

1.2.4 IMPACT SEVERITY

Climate-change impacts clearly increase with increases in the underlying climate perturbation, but how strongly? By 2012, global temperatures had increased by nearly 1 °C above pre-industrial temperatures (Jones et al 2012), equivalent to about 0.01 °C yr^{-1}, with about 20 cm sea-level rise (Church and White 2011), and there are increasing numbers of unusual weather events that have been attributed to climate change (e.g. Schneider et al 2007, Trenberth and Fasullo 2012). By the time temperature increases reach 2°, or sea-level rise reaches 40 cm, would impacts be twice as bad or increase more sharply? If impacts increase sharply with increasing perturbations, then overall damages would be largely determined by impacts at the times of highest perturbations, whereas with a less steep impact response function, impacts at times with lesser perturbations would contribute more to overall damages.

Schneider et al (2007) comprehensively reviewed and discussed the quantification of climate-change impacts and their relationship to underlying climate perturbations but concluded that a formal quantification of impacts was not yet possible. This was due to remaining scientific uncertainty, and the intertwining of scientific assessments of the likelihood of the occurrence of certain events and value judgements as to their significance.

For example, Thomas et al (2004) quantified the likelihood of species extinction under climate change and concluded that by 2050, 18% of species would be 'committed to extinction' under a low-emission scenario, which approximately doubled to 35% under a high-emission scenario. Given the functional redundancy of species in natural ecosystems, their impact on ecosystem function, and their perceived value for society, doubling the loss of species would presumably more than double the perceived impact of the loss of those species. The scientifically derived estimate of species loss therefore does not automatically translate into a usable damage response function. It requires additional value judgements, such as an assessment of the importance of the survival of species, including those without economic value.

It is also difficult to quantify the impact related to the low probability of crossing key thresholds (Lenton et al 2008). It may be possible to agree on the importance of crossing some irreversible thresholds, but it is difficult to confidently derive probabilities of crossing them. But despite these uncertainties, some kind of damage response function must be used to quantify the marginal impact of extra emission units.

As it is difficult, if not impossible, to employ purely objective means of generating impact response functions, we have to resort to what Stern called a 'subjective probability approach. It is a pragmatic response to the fact that many of the true uncertainties around climate-change policy cannot themselves be observed and quantified precisely' (Stern 2006).

Different workers have used some semi-quantitative approaches, such as polling of expert opinion (e.g. Nordhaus 1994), or the generation of complex uncertainty distributions from a limited range of existing studies (Tol 2012), but none of these overcomes the essentially subjective nature of devising impact response functions.

Figure 1 shows some possible response functions that relate an underlying climate perturbation to its resultant impact. This is quantified relative to maximum impacts anticipated over the next 100 years for perturbations such as temperature. The current temperature increase of about 1 °C is approximately 1/3 of the temperature increase expected under RCP6 over the next 100 years, giving a relative perturbation of 0.33. For the quantification of CCIPs, impacts had to be expressed as functions of relative climate perturbations to enable equal quantitative treatment of all three kinds of climatic impacts.

Economic analyses tend to employ quadratic or cubic responses function (e.g. Nordhaus 1994, Hammitt et al 1996, Roughgarden and Schneider 1999, Tol 2012), but there is concern that these functions that are based only on readily quantifiable impacts may give insufficient weight to the small probability of extremely severe impacts (e.g. Weitzman 2012, 2013, Lemoine and McJeon 2013). A response function that includes these extreme impacts would increase much more sharply than quadratic or cubic response functions (e.g. Weitzman 2012).

The relationship used here uses an exponential increase in impacts with increasing perturbations to capture the sharply increasing damages with larger temperature increases (as shown by Hammitt et al 1996 and Weitzman 2012). Warming by 3/4 of the expected maximum warming, for example, would have about 10 times the impact as warming by only 1/4 of maximum warming. The graph also shows the often-used power relationships (e.g. Hammitt et al 1996, Boucher 2012), shown here with powers of 2 (quadratic) and 3 (cubic), and a more extreme impacts function (hockey-stick function) presented by Hammitt et al (1996). Compared to the power functions, the exponential relationship calculates relatively modest impacts for moderate climate perturbations that increase more sharply for more extreme climate perturbations. It is thus very similar to the 'hockey-stick' relationship of Hammitt et al (1996).

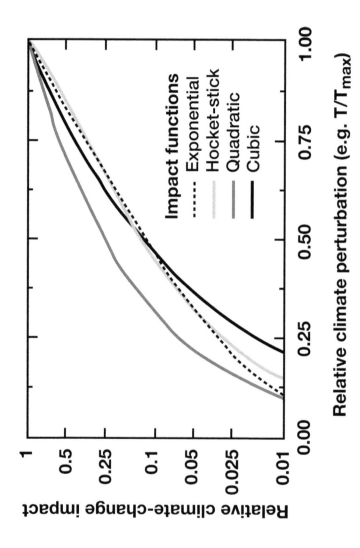

Figure 1. Quantification of climate-change impacts as a function of relative climate perturbations. This is illustrated for the exponential relationship used here, the 'hockey-stick' function of Hammitt et al (1996) and quadratic and cubic impact functions. It is shown for different relative climate perturbations, such as temperature changes, relative to the maximum perturbations anticipated over the next 100 years.

1.2.5 DISCOUNT FACTORS

Should near-term impacts be treated as more important than more distant impacts? If one applies discount rates of 4%, for example, it would render impacts occurring in just 17 years as being only half as important as impacts occurring immediately. The choice of discount rates is hence one of the most critical components of any impact analysis, and the influential Stern report (Stern 2006) derived a fairly bleak outlook on the seriousness of climate change, largely due to using an unusually low discount rate of 1.4%.

While the use of large discount rates is warranted in purely economic analyses, it is questionable in environmental assessments as it essentially treats the lives and livelihood of our children and grandchildren as less important than our own, which is hard to justify ethically (e.g. Schelling 1995, Sterner and Persson 2008). On the other hand, using a 0 discount rate would treat impacts in perpetuity as equally important as short-term impacts. This raises at least the practical problem that it becomes increasingly difficult to predict events and their significance into the more distant future.

The calculation of GWPs essentially uses 0 discount rates, but ignores impacts beyond the end of the assessment period (Tanaka et al 2010). This avoids a preferential emphasis on the impacts on one generation over another, yet avoids the unmanageable requirement of having to assess impacts in perpetuity. This approach is also used here for calculating CCIPs.

1.3 CALCULATION METHODS

1.3.1 QUANTIFYING CLIMATE-CHANGE IMPACT POTENTIALS

To quantify the three different kinds of impacts, it is necessary to first calculate the climate perturbations underlying them. The perturbation Py, T in year y, related to direct-temperature impacts, is simply calculated as:

$$P_{y,T} = T_y - T_p \tag{1}$$

where T_y and T_p are the temperatures in year y and pre-industrially. The temperature in 1900 is taken as the pre-industrial temperature.

The rate of temperature change, $P_{y,\Delta}$, is calculated as the temperature increase over a specified time frame:

$$P_{y,\Delta} = (T_y - T_{y-d})/d \tag{2}$$

where d is the length of the calculation interval, set here to 100 years. Shorter calculation intervals could be used, in principle, but extra emission units would then affect both the starting and end points for calculating rates of change, leading to complex and sometimes counter-intuitive consequences. The choice of 100 years is further discussed below.

The cumulative temperature perturbation, $P_{y,\Sigma}$, is calculated as the sum of temperatures above pre-industrial temperatures:

$$P_{y,\Sigma} = \sum_{i=p}^{y} (T_i - T_p) \tag{3}$$

where T_i is the temperature in every year i from pre-industrial times to the year y.

All three perturbations are then normalized to calculate relative perturbations, Q, as:

$$Q_{y,T} = P_{y,T}/\max(P_{T,RCP6}) \tag{4a}$$

$$Q_{y,\Delta} = P_{y,\Delta}/\max(P_{\Delta,RCP6}) \tag{4b}$$

$$Q_{y,\Sigma} = P_{y,\Sigma}/\max(P_{\Sigma,RCP6}) \tag{4c}$$

where the P-terms are the calculated perturbations under a chosen emissions pathway, and the max-terms are the maximum perturbations calculated under RCP6 over the next 100 years. With this normalization, each kind of climate impact can be treated mathematically the same.

Impacts, I, are then derived from relative perturbations as:

$$I_{y,T} = [(e^{Q_{y,T}})^s] - 1 \qquad (5a)$$

$$I_{y,\Delta} = \left[(e^{Q_{y,\Delta}})^s\right] - 1 \qquad (5b)$$

$$I_{y,\Sigma} = \left[(e^{Q_{y,\Sigma}})^s\right] - 1 \qquad (5c)$$

where s is a severity term that describes the relationship between perturbations and impacts (figure 1). The work presented here uses s = 4 (as discussed in section 2.4 above).

Temperatures from 1900 to 2010 were based on the HadCRUT4 data set of Jones et al (2012). They were used to set initial temperatures for calculating rates of warming and cumulative warming up to 2010. Temperatures beyond 2010 were added to base temperatures and together determined respective perturbations over the next 100 years.

The relevant impacts were then calculated using equation (4), and summed over 100 years. To calculate CCIPs, these calculation steps were followed four times. The first set of calculations was based on RCP6 and was only used to derive max (PRCP6) which was needed for subsequent normalizations. This normalization made it possible to assign each kind of impact equal importance at their highest perturbations over the next 100 years under RCP6.

The second set of calculations used a chosen emission pathway, RCP6, or a different one as specified below, to calculate background gas concentrations and perturbations. The final two sets of calculations used the same chosen emission pathway and added either 1 tonne of CO_2 or of a different gas. The calculations then derived marginal extra impacts of extra emission units under the three different kinds of impacts. CCIPs of each gas were then calculated as the ratios of marginal impacts of different gases relative to those of CO_2.

These calculations aim to estimate impacts over the coming 100 years, and how those impacts might be modified through pulse emissions of different GHGs. They use the best estimates of relevant background conditions based on emerging science and updated emission scenarios. These calculations would need to be repeated every few years with new scientific understanding and newer emission projections to provide updated guidance of the importance of different GHG over the next 100-year period.

1.3.2 CALCULATING RADIATIVE FORCING AND TEMPERATURE CHANGES

The calculations of radiative forcing and temperature followed the approach of Kirschbaum et al (2013), including the carbon cycle based on the Bern model and radiative forcing calculations provided by the IPCC. Calculations also included the replacement of a molecule of CO_2 by CH_4 in the biogenic production of CH_4, and its partial conversion back to CO_2 when CH_4 was oxidized (Boucher et al 2009). Global temperature calculations included a term for the thermal inertia of the climate system. Full calculation details are given in the supplemental information (available at stacks.iop.org/ERL/9/034014/mmedia).

1.4 RESULTS

1.4.1 IMPACTS UNDER BUSINESS-AS-USUAL CONCENTRATIONS

A quantification of the marginal impact of additional units of each gas must be based on background conditions that include quantification of the impacts that are expected to occur without those additional emission units. Figure 2 shows the relative perturbations underlying the three kinds of impacts and resultant calculated climate-change impacts.

Under RCP6, direct and cumulative-warming impacts continue to increase throughout the 21st century, with greatest impacts reached by 2109. Rate-of-warming impacts reach their maximum by about 2080 and then start to fall slightly (figure 2(a)). While the underlying climate perturbations increase fairly linearly over the next 100 years, this leads to sharply increasing impacts towards the end of the assessment period (figure 2(b)). This pattern is most pronounced for cumulative-warming impacts. The irregular pattern in calculated rates of warming is related to the unevenness in the observed

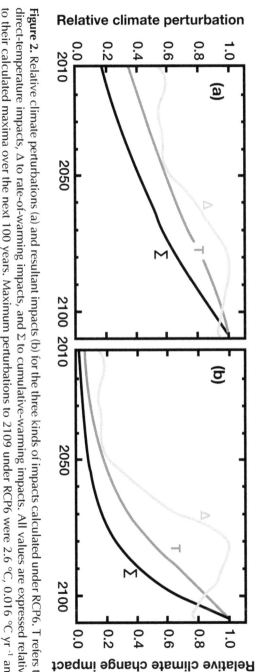

Figure 2. Relative climate perturbations (a) and resultant impacts (b) for the three kinds of impacts calculated under RCP6. T refers to direct-temperature impacts, Δ to rate-of-warming impacts, and Σ to cumulative-warming impacts. All values are expressed relative to their calculated maxima over the next 100 years. Maximum perturbations to 2109 under RCP6 were 2.6 °C, 0.016 °C yr⁻¹ and 206 °C yr, respectively.

temperature records up to 2010 as rates of warming are calculated from the temperature difference over the preceding 100 years.

1.4.2 PHYSICO-CHEMICAL EFFECTS
OF EXTRA GHG EMISSIONS

To calculate the marginal impact of pulse emissions of extra GHG units, it is necessary to first establish their physico-chemical consequences. Concentration increases are greatest immediately after the emission of extra units. They then decrease exponentially for CH_4 (figure 3(b)) and N_2O (figure 3(c)). CO_2 concentrations also decrease but follow a more complex pattern (figure 3(a)). For CH_4, the decrease is quite rapid, with a time constant of 12 years, but is more prolonged for N_2O, with a time constant of 120 years.

These concentration changes exert radiative forcing. It is also highest immediately after the emission of each gas and decreases thereafter. It decreases proportionately faster than the concentration decrease because of increasing saturation of the relevant infrared absorption bands. This is most pronounced for CO_2 (figure 3(a)), for which RCP6 projects large concentration increases (figure 3(d)), which makes the remaining CO_2 molecules from 2010 pulse emissions progressively less effective (e.g. Reisinger et al 2011). For N_2O, RCP6 projects only moderate concentration increases. The infrared absorption bands of N_2O are also less saturated than for CO_2 so that the effectiveness of any remaining molecules remains high. RCP6 projects little change in the CH_4 concentration. Radiative forcing then drives temperature changes (figure 3) that lag radiative forcing by 15–20 years due to the thermal inertia of the climate system.

1.4.3 MARGINAL IMPACTS OF EXTRA EMISSION UNITS

From the information in figures 2 and 3, one can calculate marginal increases in impacts due to a 2010 pulse emission of each gas (figure 4). Extra units of CO_2 emitted in 2010 cause the largest temperature increase in about 2025 (figure 3(a)). Base temperatures, however, are still fairly mild in 2025 (figure

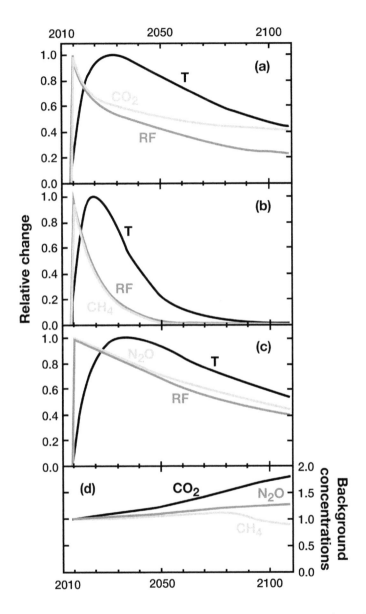

Figure 3. Calculated increases in the concentrations of CO_2 (a), CH_4 (b) and N_2O (c) due to pulse emissions of each gas in 2010 and the resultant radiative forcing and temperature increases. Also shown are relative changes in background gas concentrations according to RCP6 (d). All values in (a)–(c) are expressed relative to their highest values over the next 100 years, and concentrations in (d) are relative to 2010 concentrations.

2(a)) so that the extra warming at that time increases direct-temperature impacts only moderately (figure 4(a)). Even though the extra warming from CO_2 added in 2010 diminishes over time (figure 3(a)), it adds to increasing base temperatures (figure 2(a)) to cause increasing ultimate impacts (figure 4(a)). This pattern is strongest for cumulative-warming impacts. The patterns for N_2O (figure 4(c)) are similar to those for CO_2 because the longevity of N_2O in the atmosphere is similar to that of CO_2.

CH$_4$ emitted in 2010, however, modifies direct-temperature impacts only over the first few decades after its emission (figure 3(b)). While later warming could potentially have greater impacts, the residual warming several decades after its emission becomes so small to have very little effect. For cumulative-warming impacts, however, the greatest marginal impact of CH_4 additions also occurs at the end of the assessment period. Even though CH_4 emissions exert their warming early in the 21st century, that warming is effectively remembered in the cumulative temperature record and leads to the largest ultimate impact when it combines with large cumulative-warming base impacts (figure 4(b)).

For rate-of-warming impacts and direct-temperature impacts, there are distinctly different patterns for the different gases that are principally related to the longevity of the gases in the atmosphere. For cumulative-warming impacts, however, the patterns are similar for all gases, with the marginal impact from a 2010 pulse emission being muted for the first 50–80 years and then increasing sharply over the remainder of the 100-year assessment period. This is because cumulative warming can be increased in much the same way for contributions made earlier as from on-going warming. Even though different gases make their additions to cumulative warming at different times, that increased perturbation has the largest impact when it adds to large base values (see figure 2(a)) so that for all gases, the largest marginal impacts occur at the end of the 100-year assessment period (figure 4).

1.4.4 CLIMATE-CHANGE IMPACT POTENTIALS

Marginal impacts can then be summed over the 100 years after the pulse emission of each gas and expressed relative to CO_2 (table 1). Under RCP6,

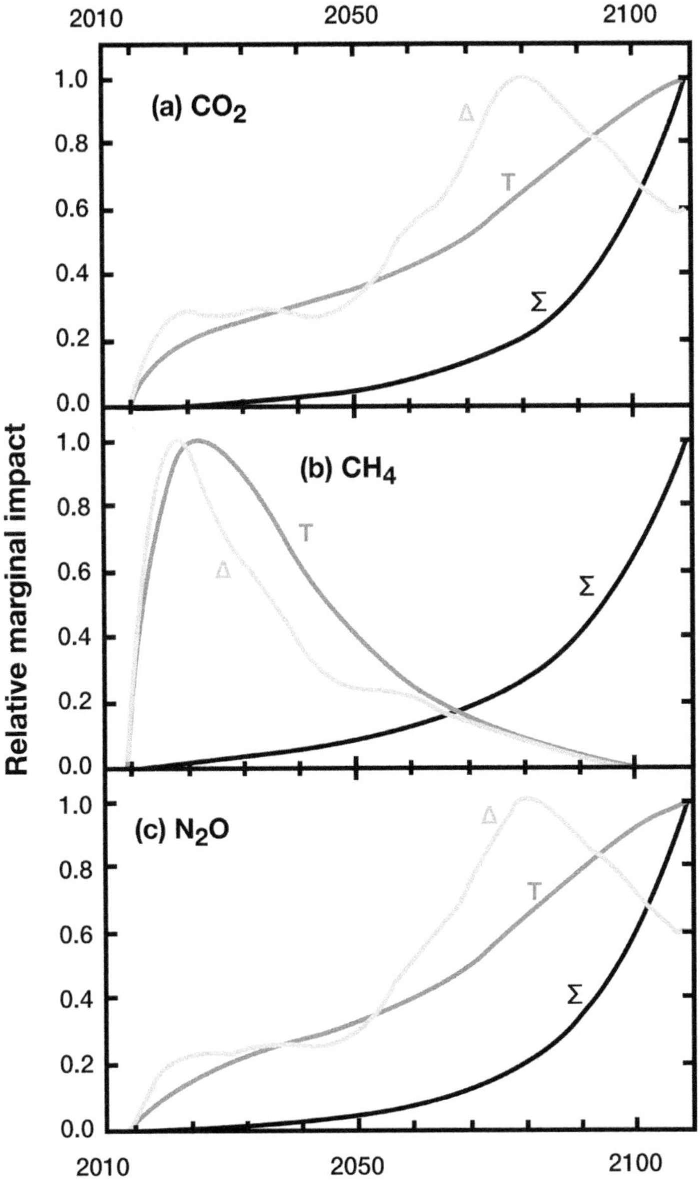

Figure 4. Change in the three kinds of impacts due to the addition of one unit of CO_2 (a), biogenic CH_4 (b) and N_2O (c) in 2010. Symbols as for figure 2. All numbers are normalized to the highest marginal impacts calculated over the next 100 years.

CCIPs for biogenic and fossil CH_4 are 20 and 23, respectively, compared to a 100-year GWP of 25. These lower values are due to the much lower direct-temperature and rate-of-warming impacts. Peak warming from CH_4 emissions in 2010 occurs at a time when background temperature increases are still fairly mild so that the extra warming from CH_4 (figure 4(b)) causes less severe impacts than the warming from CO_2 (figure 4(a)) that is still strong many decades later when it combines with higher background temperatures to cause more severe additional impacts.

In contrast, cumulative-warming impacts under RCP6 are 34 and 37 (for biogenic and fossil CH_4), which are greater than the corresponding values for cumulative radiative forcing. The earlier radiative forcing from CH_4 ensures that all radiative forcing leads to warming within the assessment period. For CO_2 and N_2O, on the other hand, radiative forcing overestimates their warming impact because of the thermal inertia of the climate system. Some of the radiative forcing exerted towards the end of the 100-year assessment period only leads to warming after the end of the assessment period providing relatively more cumulative radiative forcing than cumulative warming.

Fossil-fuel-derived CH_4 has higher CCIPs than biogenic CH_4 by about three units. Biogenic CH_4 production means that a molecule of carbon is converted to CH_4, which lowers the atmospheric CO_2 concentration and thereby reduces its overall climatic impact. After it has been oxidized, any CH_4, however, continues its radiative forcing as CO_2, which increases its overall impact (Boucher et al 2009), with the same effect for both fossil and biogenic CH_4.

CCIPs of CH_4 become progressively smaller when they are calculated under higher concentration pathways (table 1). This is caused by much higher impact damages being reached under higher concentration pathways so that the earlier warming contribution of CH_4 relative to CO_2 becomes increasingly less important. This affects direct-temperature impacts and rate-of-warming impacts, whereas cumulative temperature impacts remain similar under the different RCPs.

For N_2O, the CCIP is greater than the 100-year GWP (348 versus 298 under RCP6). This is mainly due to the reducing effectiveness of infrared absorption of extra CO_2 under increasing background concentrations, which increases the relative importance of the emission of other gases. This

Table 1. Cumulative radiative forcing, the three kinds of impacts calculated separately and combined to calculate CCIPs over 100 years. Calculations are done under constant 2010 GHG concentrations, and under four different RCPs. All numbers are expressed relative to CO_2. Calculations are done separately for biogenic (B) and fossil-derived (F) CH_4. CCIPs are calculated as the average of the three individually calculated kinds of impacts. Calculations under RCP6 are shown in bold as the reference condition used here. Constant 2010 concentrations were taken to be 387, 1.767 and 0.322 ppmv for CO_2, CH_4 and N_2O, respectively. Numbers for cumulative radiative forcing are given only for comparison. Currently used 100-year GWPs are 25 for CH_4 and 298 for N_2O.

		Cumulative radiative forcing	Direct-temperature impacts	Rate-of-warming impacts	Cumulative-warming impacts	CCIPs
CH_4 (B)	Const	22	23	34	32	29
CH_4 (F)		24	25	36	34	32
N_2O		282	285	285	288	286
CH_4 (B)	RCP3	24	24	32	35	30
CH_4 (F)		27	27	35	37	33
N_2O		306	313	313	313	313
CH_4 (B)	RCP4.5	26	16	19	34	23
CH_4 (F)		29	19	22	37	26
N_2O		331	341	342	328	337
CH_4 (B)	RCP6	27	12	13	34	20
CH_4 (F)		30	15	16	37	23
N_2O		338	359	356	329	348
CH_4 (B)	RCP8.5	29	5.0	3.9	34	14
CH_4 (F)		32	7.8	6.7	37	17
N_2O		365	437	438	351	408

interaction with base-level gas concentrations is not included in GWPs as they are calculated under constant background gas concentrations.

1.4.5 THE IMPORTANCE OF CLIMATE-CHANGE SEVERITY

The relative importance of different gases also depends strongly on the underlying climate-change severity term (figure 5). With increases in the

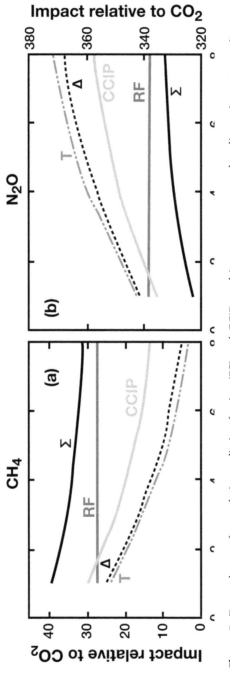

Figure 5. Dependence of cumulative radiative forcing (RF) and CCIPs and its components on the climate-impact severity terms for biogenic CH_4 (a) and N_2O (b). T refers to direct-temperature impacts, Δ to rate-of-warming impacts, Σ to cumulative-warming impacts, RF to radiative forcing and CCIP to the derived combined index.

severity term, the importance of short-lived CH_4 decreases considerably (figure 5(b)), whereas the importance of N_2O increases slightly (figure 5(a)). This is because the greatest temperature and rate of change perturbations are projected to occur at the end of the assessment period when CH_4 adds little to those perturbations, while N_2O adds even more than CO_2. As the climate-change severity term increases, it progressively increases the importance of impacts at these later periods and thereby greatly reduces the importance of CH_4.

1.5 DISCUSSION

In this work, climate-change impact potentials are presented as an alternative metric for comparing different GHGs. Why use a new metric? The ultimate aim of climate-change mitigation is to avert adverse climate-change impacts. Hence, there is an obvious logic for policy setting to start with a clear definition of the different kinds of climatic impacts that are to be avoided. Climate-change metrics are needed to support that climate-change policy with the same definition and quantification of climate-change impacts so that the effects of different GHGs can be compared. Mitigation efforts can then be targeted at the gases through which mitigation efforts can be achieved most cost-effectively.

The key aim of metrics should be the quantification of the marginal impact of pulse emissions of extra GHG units. CCIPs aim to provide that measure. They aim to achieve that by combining simple calculations of the relevant physics and atmospheric chemistry with an assessment of the key impacts on nature and society. This full assessment is needed to underpin the development of the most cost-effective mitigation strategies.

The calculation of CCIPs begins with setting the most likely background conditions with respect to gas concentrations and background temperatures in order to quantify the marginal impact of an extra emission unit of a GHG. The use of CCIPs thus requires a periodic re-evaluation of background conditions to devise new optimal mitigation strategies. It is necessary for mitigation efforts to be continuously refocused to achieve the most cost-effective climate-change impact amelioration (Johansson et al 2006). This is first because the relative importance of extra GHG units diminishes with an increase in their own background concentrations because of increasing saturation of the relevant infrared absorption bands. Conversely, the extra warming caused by additional emission units has a greater impact when background temperatures are already higher as it can contribute towards raising temperatures into an increasingly dangerous range. The marginal impact of extra emission units can, therefore, only be quantified under a specified emissions pathway and time horizon.

Along the chain of causality from greenhouse gas emissions to ultimate climate-change impacts, the relevance of respective metrics increases but the uncertainty associated with their calculation increases as well (e.g., Fuglestvedt et al 2003). This relates to scientific uncertainty, the value judgements needed about the relative importance of different impacts, and the ethical considerations of accounting for impacts occurring at different times. GWPs are at one extreme of this continuum, requiring a minimum of assumptions in their calculation, but they only quantify a precursor of ultimate impacts. CCIPs try to go several steps further by quantifying specific climate perturbations that are more directly related to different kinds of climate-change impacts. The functions used to calculate CCIPs still retain simplicity and transparency.

The use of CCIPs instead of 100-year GWPs would reduce the short-term emphasis on CH_4 as CH_4 emitted in 2010 will have disappeared from the atmosphere by the time the most damaging temperatures or rates of warming will be reached. This conclusion is similar to that reached by studies based on GTP calculations (e.g. Shine et al 2007). However, even CH_4 contributes to cumulative-warming impacts. Using CCIPs would thus make CH_4 less important without rendering it irrelevant. Over time, and if future GHG emissions remain high, CH_4 is likely to become more important as the time of emission gets closer to the times when the most severe impacts may be anticipated (Shine et al 2007, Smith et al 2012). CH_4 would then increasingly contribute not only to cumulative-warming impacts but also to direct-temperature and rate-of-warming impacts. CCIPs would need to be recalculated periodically in line with continuously changing expectations of the future.

CCIPs also change with background conditions, and it is considered likely that CCIPs calculated under RCP6 are the most relevant. Recent concentration trends, even during times of global economic crisis (e.g.

Peters et al 2013), point towards higher concentration pathways. The limited willingness of the international community to seriously address climate change also suggests that higher concentration pathways will be more likely. RCP6 was therefore used here as the most likely background condition from which to assess the marginal impacts of the emission of extra GHG units.

The derivation of CCIPs explicitly defined and quantified three distinct kinds of impacts. They were all related to temperature as even climate impacts such as flooding that may be more directly related to rainfall intensity can be related to temperature-based perturbations as the underlying climatic driver. One impact that cannot be related to temperature is the direct impact of elevated CO_2 itself. Increasing CO_2 leads to ocean acidification (e.g. Kiessling and Simpson 2011) and shifts the ecological balance between plant species, especially benefiting C_3 plants at the expense of C_4 plants (e.g. Galy et al 2008). On the other hand, increasing CO_2 is beneficial through increasing biological productivity and may partly negate the pressures on food production from increasing temperatures and precipitation changes (e.g. Jaggard et al 2010). With these divergent impacts its overall net impact remains uncertain, and may even be regarded as either positive or negative, and it was, therefore, not included in the CCIP calculations.

Another critically important factor is the steepness of the relationship between underlying climate perturbations and their resultant impact (figures 1, 5). The more steeply impacts increase with increasing perturbation, the more it shifts the importance of extra warming to the times when impacts are already high. With a less steep impact curve, warming at times with lower background temperatures also makes significant contributions towards the overall impact load. The present work used a response function similar to the 'hockey-stick' function first presented by Hammitt et al (1996). This steep response curve emphasizes the contribution of different gases at the times of highest impacts while reducing the importance of their contributions at times of lower background impacts. It thus reduces the importance of short-lived gases such as CH_4.

It is also important over what time interval relevant rate-of-warming impacts are calculated. The present work used an assessment period of 100 years and calculated the rate of warming from the temperature increase over the preceding 100 years. With these assumptions, the starting temperatures are always part of the immutable past and extra emissions affect only the end-point temperatures. However, a calculation interval of 100 years may be regarded as too long (e.g. Peck and Teisberg 1994) and a shorter interval might be seen as more appropriate.

Shortening the calculation interval to less than 100 years, however, creates complex interactions because extra emission units then affect both the starting and end point for calculating the rate of warming, and results can become complex and counter-intuitive. For instance, if rates of warming were calculated over 50 years, extra methane emissions would paradoxically reduce the sum of calculated rate-of-warming impacts. How does that occur? While extra methane would increase temperatures over a few decades after its emission, it would increase the short-term rate of warming (calculated from, say, 1970–2020), but it would reduce the rate of warming calculated from 2020 to 2070. Since higher rates of warming are anticipated later during the 21st century (see figure 2), the gains from decreasing the more damaging rates of warming at the later time would be greater than the harm from marginally worsening the milder rates-of-warming impacts in the shorter term.

Whether extra emissions would be considered to do ultimate harm or good would thus depend on the timing of those respective increased and decreased perturbations relative to the base perturbations and the length of period that is assessed as most appropriate for assessing rate-of-warming impacts. Exploring these complex interactions is beyond the scope of the present paper, and the present work had to restrict itself to the simpler case of calculating rates-of-warming impacts by the temperature change over 100 years.

The calculation of CCIPs cannot be based purely on objective science, but has to combine scientific insights with value judgements and assumptions about future background conditions. They relate to the steepness of the impacts function, the choice of background scenario, the inclusion or exclusion of time discounting, the length of assessment horizon, the relative weighting assigned to the different kinds of impacts, the length of the time period for quantifying the rate of change and others. These choices all have a bearing on calculated CCIPs. It may be seen as unfortunate that CCIPs cannot be developed without recourse to a number of key assumptions. However, society makes these assumptions implicitly whenever it decides on adopting any policies related to climate change. The process that is followed formally and explicitly in this paper is similar to the process followed implicitly in all discussions of the importance of climate change, and that has led to the current level of concern and partial willingness to pursue mitigative measures.

Various possible metrics to account for different GHG emissions have been proposed in the past (Ekholm et al 2013). They fall under three broad categories: (1) using measures of cumulative radiative forcing, such as for GWPs; (2) rate of warming, like that explored by Peck and Teisberg (1994); and (3) a number of proposals that are predicated on impacts related directly to elevated temperature, such as the Global Damage Potential (Kandlikar 1996, Hammitt et al 1996, see also Boucher 2012), and the Global Temperature Change Potential, GTP, (Shine et al 2005, 2007). The present work is the first to derive a metric explicitly based on all three kinds of impacts.

Metrics may also restrict themselves to the use of physico-chemical quantities, such as the GWP or the GTP, or employ detailed economic analyses to derive ultimate cost or damage functions (e.g. Kandlikar 1996, Manne and Richels 2001). Including explicit models to calculate damages aims to get closer to an explicit calculation of the ultimate impacts that matter, but it greatly reduces the transparency of resultant metrics (Johansson 2012). It also tends to bias analyses towards those aspects that can be quantified more readily, such as economic impacts, while other impacts, such as the perceived loss from the extinction of species, or the damage from low-probability, but high impact, events tend to be ignored (e.g. Weitzman 2012). The present work restricts itself to using simple models for calculating physico-chemical processes that allowed a number of critical assumptions to remain explicit and transparent. It thereby aims to retain the transparency needed for adoption in international policy or research applications.

1.6 CONCLUSIONS

Global Warming Potentials calculated over 100 years are the current default metric to compare different GHGs. They have become the default metric despite the recognition that they are not directly related to the ultimate climate-change impacts that society is trying to avert. To achieve mitigation objectives most cost-effectively, and to be able to target an optimal mix of GHGs, requires a clearer definition of what is to be avoided. This, in turn, necessitates a more complex analysis than provided by the use of GWPs.

Over the years, there have been several proposals of alternative accounting metrics. A key difference between these different metrics lies in their damage functions that may be related directly to elevated temperature (e.g. Kandlikar 1996, Shine et al 2005, 2007), or to the rate of warming (Peck and Teisberg 1994), or to a measure of cumulative radiative forcing (as for GWPs). However, no previously proposed metric explicitly included all three different kinds of climate perturbations that all contribute towards overall impacts (e.g. Fuglestvedt et al 2003, Brandão et al 2013). Instead, previous work derived respective metrics based on only one of these kinds of impacts and thus implicitly negated the importance of the other kinds of impacts. CCIPs are the first attempt to explicitly develop a metric that is based on all three kinds of impacts.

Climate change continues to be a significant threat for the future of humanity, and mitigation is needed to avert those threats as much as possible. The global community, however, is showing only a limited willingness to allocate sufficient resources to avert serious long-term impacts. The development of CCIPs aims to assist in using those limited resources as cost-effectively as possible.

REFERENCES

1. Baker A C, Glynn P W and Riegl B 2008 Climate change and coral reef bleaching: an ecological assessment of long-term impacts, recovery trends and future outlook Estuar. Coast. Mar. Sci. 80 435–71
2. Boucher O 2012 Comparison of physically- and economically-based CO2-equivalences for methane Earth Syst. Dyn. 3 49–61
3. Boucher O, Friedlingstein P, Collins B and Shine K P 2009 The indirect global warming potential and global temperature change potential due to methane oxidation Environ. Res. Lett. 4 044007
4. Brandão M et al 2013 Key issues and options in accounting for carbon sequestration and temporary storage in life cycle assessment and carbon footprinting Int. J. Life Cycle Assess. 18 230–40

5. Church J A and White N J 2011 Sea-level rise from the late 19th to the early 21st century Surv. Geophys. 32 585–602

6. Ekholm T, Lindroos T J and Savolainen I 2013 Robustness of climate metrics under climate policy ambiguity Environ. Sci. Policy 31 44–52

7. Fuglestvedt J S, Berntsen T K, Godal O, Sausen R, Shine K P and Skodvin T 2003 Metrics of climate change: assessing radiative forcing and emission indices Clim. Change 58 267–331

8. Fuglestvedt J S et al 2010 Transport impacts on atmosphere and climate: metrics Atmos. Environ. 44 4648–77

9. Galy V, Francois L, France-Lanord C, Faure P, Kudrass H, Palhol F and Singh S K 2008 C4 plants decline in the Himalayan basin since the Last Glacial Maximum Quat. Sci. Rev. 27 1396–409

10. Hammitt J K, Jain A K, Adams J L and Wuebbles D J 1996 A welfare based index for assessing environmental effects of greenhouse-gas emissions Nature 381 301–3

11. Huang C R, Barnett A G, Wang X M, Vaneckova P, FitzGerald G and Tong S L 2011 Projecting future heat-related mortality under climate change scenarios: a systematic review Environ. Health Perspect. 119 1681–90

12. Hughes L, Cawsey E M and Westoby M 1996 Climatic range sizes of Eucalyptus species in relation to future climate change Glob. Ecol. Biog. Lett. 5 23–9

13. Jaggard K W, Qi A M and Ober E S 2010 Possible changes to arable crop yields by 2050 Phil. Trans. R. Soc. B 365 2835–51

14. Johansson D J A 2012 Economics- and physical-based metrics for comparing greenhouse gases Clim. Change 110 123–41

15. Johansson D J A, Persson U M and Azar C 2006 The cost of using global warming potentials: analysing the trade off between CO2, CH4 and N2O Clim. Change 77 291–309

16. Jones P D, Lister D H, Osborn T J, Harpham C, Salmon M and Morice C P 2012 Hemispheric and large-scale land-surface air temperature variations: an extensive revision and an update to 2010 J. Geophys. Res. 117 D05127

17. Kandlikar M 1996 Indices for comparing greenhouse gas emissions: integrating science and economics Energy Econ. 18 265–81

18. Kendall A 2012 Time-adjusted global warming potentials for LCA and carbon footprints Int. J. Life Cycle Assess. 17 1042–9

19. Kiessling W and Simpson C 2011 On the potential for ocean acidification to be a general cause of ancient reef crises Glob. Change Biol. 17 56–67

20. Kirschbaum M U F 2003a Can trees buy time? An assessment of the role of vegetation sinks as part of the global carbon cycle Clim. Change 58 47–71

21. Kirschbaum M U F 2003b To sink or burn? A discussion of the potential contributions of forests to greenhouse gas balances through storing carbon or providing biofuels Biomass Bioenergy 24 297–310

22. Kirschbaum M U F 2006 Temporary carbon sequestration cannot prevent climate change Mitig. Adapt. Strateg. Glob. Change 11 1151–64

23. Kirschbaum M U F, Saggar S, Tate K R, Thakur K and Giltrap D 2013 Quantifying the climate-change consequences of shifting land use between forest and agriculture Sci. Tot. Environ. 465 314–24

24. Lashof D A and Ahuja D R 1990 Relative contributions of greenhouse gas emissions to global warming Nature 344 529–31

25. Lemoine D and McJeon H C 2013 Trapped between two tails: trading off scientific uncertainties via climate targets Environ. Res. Lett. 8 034019

26. Lenton T M, Held H, Kriegler E, Hall J W, Lucht W, Rahmstorf S and Schellnhuber H J 2008 Tipping elements in the Earth's climate system Proc. Natl Acad. Sci. USA 105 1786–93

27. Manne A S and Richels R G 2001 An alternative approach to establishing trade-offs among greenhouse gases Nature 410 675–7

28. Manning M and Reisinger A 2011 Broader perspectives for comparing different greenhouse gases Phil. Trans. R. Soc. A 369 1891–905

29. Meehl G A et al 2012 Relative outcomes of climate change mitigation related to global temperature versus sea-level rise Nature Clim. Change 2 576–80

30. Nordhaus W D 1994 Expert opinion on climatic-change Am. Sci. 82 45–51

31. Parmesan C and Yohe G 2003 A globally coherent fingerprint of climate change impacts across natural systems Nature 421 37–42

32. Peck S C and Teisberg T J 1994 Optimal carbon emissions trajectories when damages depend on the rate or level of global warming Clim. Change 28 289–314

33. Peters G P, Aamaas B, Berntsen T and Fuglestvedt J S 2011a The integrated global temperature change potential (iGTP) and relationships between emission metrics Environ. Res. Lett. 6 044021

34. Peters G P, Aamaas B, Lund M T, Solli C and Fuglestvedt J S 2011b Alternative 'Global Warming' metrics in life cycle assessment: a case study with existing transportation data Environ. Sci. Technol. 45 8633–41

35. Peters G P et al 2013 The challenge to keep global warming below 2 °C Nature Clim. Change 3 4–6

36. Plattner G K, Stocker P, Midgley P and Tignor M 2009 IPCC Expert Meeting On the Science of Alternative Metrics (Available at: www.ipcc.ch/pdf/supporting-material/expert-meeting-metrics-oslo.pdf)

37. Reisinger A, Meinshausen M and Manning M 2011 Future changes in global warming potentials under representative concentration pathways Environ. Res. Lett. 6 024020

38. Roughgarden T and Schneider S H 1999 Climate change policy: quantifying uncertainties for damages and optimal carbon taxes Energy Policy 27 415–29

39. Schelling T C 1995 Intergenerational discounting Energy Policy 23 395–401

40. Schneider S H et al 2007 Assessing key vulnerabilities and the risk from climate change Climate Change 2007: Impacts, Adaptation and Vulnerability. Contribution of WGII to the Fourth Assessment Report of the IPCC ed M L Parry, O F Canziani, J P Palutikof, P J van der Linden and C E Hanson (Cambridge: Cambridge University Press) 779–810

41. Shine K P, Berntsen T K, Fuglestvedt J S, Skeie R B and Stuber N 2007 Comparing the climate effect of emissions of short- and long-lived climate agents Phil. Trans. R. Soc. A 365 1903–14

42. Shine K P, Fuglestvedt J S, Hailemariam K and Stuber N 2005 Alternatives to the global warming potential for comparing climate impacts of emissions of greenhouse gases Clim. Change 68 281–302

43. Smith S M, Lowe J A, Bowerman N H A, Gohar L K, Huntingford C and Allen M R 2012 Equivalence of greenhouse-gas emissions for peak temperature limits Nature Clim. Change 2 535–8

44. Stern N H 2006 The Economics of Climate Change (Available at www.hmtreasury. gov.uk/independent_reviews/stern_review_economics_climate_change/stern_ review_report.cfm)

45. Sterner T and Persson U M 2008 An even sterner review: introducing relative prices into the discounting debate Rev. Environ. Econ. Policy 2 61–76

46. Tanaka K, Peters G P and Fuglestvedt J S 2010 Policy update: multicomponent climate policy: why do emission metrics matter? Carbon Manag. 1 191–7

47. Thomas C D et al 2004 Extinction risk from climate change Nature 427 145–8

48. Tol R S J 2012 On the uncertainty about the total economic impact of climate change Environ. Resour. Econ. 53 97–116

49. Trenberth K E and Fasullo J T 2012 Climate extremes and climate change: the Russian heat wave and other climate extremes of 2010 J. Geophys. Res.-Atmos. 117 D17103

50. van Vuuren D P et al 2011 The representative concentration pathways: an overview Clim. Change 109 5–31

51. Vermeer M and Rahmstorf S 2009 Global sea level linked to global temperature Proc. Natl Acad. Sci. USA 106 21527–32

52. Webster P J, Holland G J, Curry J A and Chang H R 2005 Changes in tropical cyclone number, duration, and intensity in a warming environment Science 309 1844–6

53. Weitzman M L 2012 GHG targets as insurance against catastrophic climate damages J. Public Econ. Theory 14 221–44

54. Weitzman M L 2013 A precautionary tale of uncertain tail fattening Environ. Resour. Econ. 55 159–73.

PART II

REDUCING CARBON FOOTPRINT

Sustainable Development and Technological Impact on CO₂ Reducing Conditions in Romania

LUCIAN-IONEL CIOCA, LARISA IVASCU, ELENA CRISTINA RADA, VINCENZO TORRETTA, AND GABRIELA IONESCU

2.1 INTRODUCTION

Sustainable development is a major concern as countries attempt to implement strategies to reduce greenhouse gas emissions. This involves reducing environmental pollution and the use of resources, eventually contributing to the welfare of society, i.e., to the improvement of the social conditions of the population. Sustainable development relies on the contribution each country worldwide. This encourages creative thinking in putting forward strategies and in the planning and development of cities and communities. The idea of sustainable development materialized at the Rio Summit. Agenda 21 was established in Rio in 1992, which integrates the principles and imperatives worldwide for sustainable development [1]. As shown by Häikiö [2], sustainable development implies the idea of a situation/world which is better than the current one, the direction towards a society based on reuse and reduction.

Because there is a need to develop a benchmark for assessing sustainable development, the European Commission substantiates the requirements of this concept. In 2000, the European Commission launched the tool of "triple base line" based on the requirements of sustainable development. It integrates three major subordinates:

- Social: organizational impact on employees, community, customers, suppliers, stakeholders, the public, media;
- Economic: financial performance development based on the principles and laws in business;
- Environment: it refers to the organization's impact on the environment through the processes and activities that they perform [3].

In the context of real energy challenges, both in terms of sustainable use of resources and emissions of CO_2 and the security of energy supply, Romania has achieved a balance in this regard. Climate change affects all of Europe, with a wide range of effects on society and the environment. Other impacts are expected in the future that may lead to significant damage, according to the most recent estimates published by the European Environment Agency. "Climate changes are a reality throughout the world, and their scale and rapidity are becoming more evident. This means that each component of the economy, including households, must adapt and reduce emissions—Jacqueline McGlade (Executive Director of the European Environment Agency) [4]".

Extreme weather phenomena in most regions, such as climate change, heat waves, flooding and droughts have triggered, in recent years, increased environmental damage throughout Europe. Although more evidence is required to discern the role of anthropic activities in climate change, increasing human activity in areas prone to hazards has been a key factor. It is expected that future climate change will increase this vulnerability because extreme weather events may become more intense and frequent. Romania's National Strategy on Climate Change 2013–2020 (SNSC) aims at reducing emissions of greenhouse gases and adapting to the inevitable negative effects of climate change on natural and anthropic systems. Greenhouse gas emissions are those stipulated in the Kyoto Protocol of the six gases responsible for the greenhouse effect. This indicator measures the greenhouse gas emissions: carbon dioxide (CO_2), nitrous oxide (N_2O), methane (CH_4) and three halocarbons (hydrofluorocarbons-HFCs, perfluorocarbons-PFCs, sulphur hexafluoride-SF6) measured by the global warming potential [5]. Of all the items on the list, CO_2 is considered to be the most significant contributor to climate change. Each of the six gases listed above has its own global warming potential based on its radioactive capacity compared to CO_2. Second and third, in terms of importance, are CH_4 and N_2O with a considerable contribution to global warming and environmental change.

Figure 1a shows the total emissions of greenhouse gases in Romania, including land use, land-use change and forestry (LULUCF), CO_2 equivalent, and the total emission of greenhouse gases, but excluding LULUCF CO_2 equivalent [5] between 2000 and 2011. The analysis of these values shows that the total GHG emission level has had a downward evolution since 2006 in Romania, which has been an improvement. Figure 1b shows the situation of Romania's GHG emission by sector of activity. The energy sector is thus developing the largest quantity of GHG per year, with a slight decrease since 2006. In Romania, the data related to the years 2000–2011 are presented by the National Institute of Statistics (NIS). For the years 2012–2014, NIS does not present the data on GHG emissions.

This study is based on the systematization of information on the directives of the European Commission concerning the reduction in the use of natural resources, the processing of sets of relevant statistical indicators, and the analysis and identification of solutions suitable to mitigate the impact of these carbon dioxide emissions through use of renewable resources. Figure 1b shows that GHG started to decline in 2009 because all organizations in all activity sectors began using renewable resources and began to take on board the concept of sustainable development. The emission level decreased as a result of economic instability which had impacted transport, various branches of production and other related activities. Romania was also affected by economic instability [6]. Since 2011, Romania has experienced economic growth, with an increase in labour productivity/employment (a growth rate of 7.1% in 2006, 5.9%

in 2007, 5.3% in 2008, and decreased rate of 4.7% in 2009, 0.9% in 2010 and a growth rate of 2.9% in 2011). The Romanian Government supports the use of renewable sources through several legal norms: the green certificates system (the producer receives from the national energy organization a number of free green certificates for the energy it produces and delivers to the network) and the subsidies granted to the producers and users of renewable sources. This sustainable behaviour in Romania is aligned with EU requirements on the use of renewable resources.

2.2 CO$_2$ EMISSIONS IN ROMANIA AND THE EU TODAY

Scientific studies [7,8] show that the greatest amount of CO$_2$ emissions, in the European Union, results from the production of electricity and heat (for example, the production of coal-based energy in the EU Member States generated approximately 950 million tons of CO2 emissions in 2005, equivalent to 24% of total CO$_2$ emissions in the EU). Cities and urban agglomerations in each country have an essential role to play in mitigating climate change, given the fact that they consume three-quarters of the energy produced in the EU and are responsible for a similar percentage of CO$_2$ emissions.

Under the Kyoto Protocol, Romania was obligated to reduce emissions of greenhouse gases by 8% between 2008 and 2012 as compared with 1989. Romania's objective is to double its energy production to about 100 TWh before 2020 [5]. Moreover, there also exists the International Municipalities Convention signed by 6160 cities [9]. This Convention is one of the most important European actions involving regional and local authorities that commit themselves voluntarily to increase energy efficiency and the use of renewable energy sources in their territories. Through their commitment, signatories of the Convention aim at achieving and surpassing the EU objective of reducing CO$_2$ emissions by 20% before 2020, in every city which adhered to the Convention. In Romania, 83 towns/cities from a total of 320 cities have acceded to this Convention. Across Europe there are 6149 signatories and 11 signatories in Asia. Romania is constantly trying to reduce environmental pollution by reducing greenhouse gas emissions.

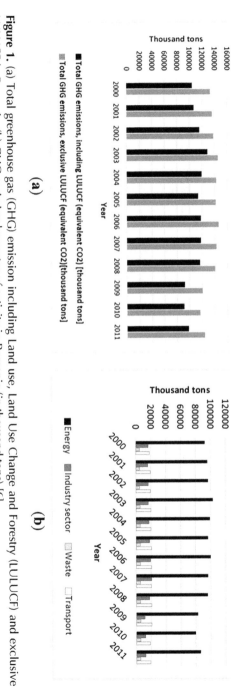

Figure 1. (a) Total greenhouse gas (GHG) emission including Land use, Land Use Change and Forestry (LULUCF) and exclusive LULUCF in Romania; (b) GHG emission by sector of activity in Romania (in thousand tons) [6].

The political commitment assumed by signing the Covenant of Mayors is transposed into the development of directions and strategies for attaining the 20% reduction in CO_2 emissions, and the use of renewable energy sources.

By analysing the amount of CO_2 in the European Union, it was found that the transportation sector is the second heaviest polluter, after the electrical and thermal energy sector. The graphical representation is showed in Figure 2a for the situation in the European Union and in Figure 2b for the situation in Romania. As shown in Figure 2, Romania's situation has improved by 1.2%, as progress has been made in reducing emissions from the transportation sector. This improvement was the effect of introducing an expensive tax for cars non-compliant with the Euro4 emission standard (the Euro4 emission standard specifies a maximum limit of 25 mg/kg particulate matters (PM) and 250 mg/kg of nitrogen oxides (NO_x)). Thus, the population shifted their preference towards vehicles compliant with the Euro5 and Euro6 standards. The rules related to the pollution standard are set by the regulation of the European Parliament and Council on the approval of road vehicles. The Euro5 emission standard requires reducing emissions by 80% compared to the Euro4 emission standard, allowing 5 mg/km of PM and 160–180 mg/km of NO_x. The Euro6 emission standard reduces the values of PM and NO_x, in comparison with Euro5, to 1 mg/km of PM and 40 mg/km of NO_x.

Figure 2a,b shows that CO_2 emissions are produced mainly by electricity and heat production, transportation, and the industrial sector. For each sector, the situation in Romania is presented and the renewable resources that can be used to alleviate the current situation are evaluated. The waste sector is also represented as well as Romania's involvement in recycling waste in comparison with the EU.

2.2.1 THE ENERGY SECTOR

The energy sector is the basis of development of a country as part of the economic infrastructure. In the current context, sustainable development involves meeting the energy demand, not by increasing energy supply

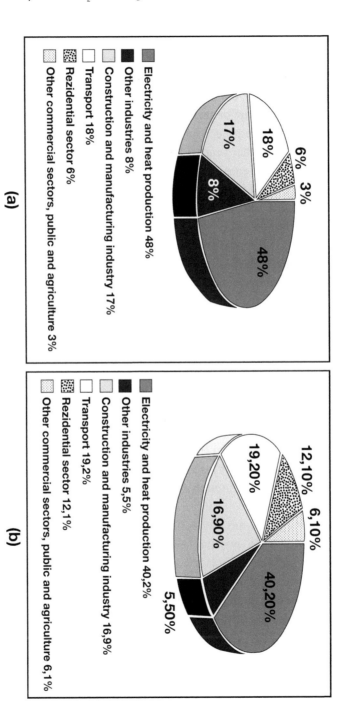

Figure 2. (a) The CO_2 emissions of different business sectors in the EU. (b) The CO_2 emissions of different business sectors in Romania.

(except for the provision of renewable energy), but through reducing energy consumption by improving technologies, by restructuring the economy and by changing attitudes concerning the efficient use of energy. The intensity of CO_2 emissions between 2000 and 2011 is shown in Figure 3. The intensity value for each year was calculated as the ratio between CO_2 emissions from energy consumption in the year and the Gross Domestic Product (GDP) in that year.

According to Figure 3, the reduction of CO_2 impact on the environment for the energy sector was achieved through the use of renewable energy sources (RES). These sources include forms of energy derived from renewable, natural processes, in which the production cycle takes place in periods that are directly proportional to their consumption periods. Thus, the energy of sunlight, winds, flowing water, biomass and geothermal heat may be captured by using different technologies [10,11,12,13]. Renewable energy is extremely important today, being considered to play a crucial role in increasing the security of energy resources by reducing dependence on fossil fuels and reducing greenhouse gas emissions. Several researchers have focused on the analysis of renewable energy resources in the European Union [14,15,16,17,18]. The proposed RES measures have brought considerable improvement, but have not shown evidence of full involvement in sustainable development.

The consumption of energy from RES in the year 2011 and target in 2020 for Romania and for EU-29 (European Union with 27 members plus Switzerland and Norway) is shown in Figure 4. In 2011, the RES share in the final energy consumption of the EU was 13.0% compared to 8.5% in 2005 [19].

In recent years, the production of renewable energy has shifted from being viewed as a possible alternative into an obligatory alternative. Romania's potential in relation to renewable energy resources in the total electrical energy is shown in Figure 5. Renewable energy sources taken into consideration in this analysis are: bioenergy, geothermal energy, wind energy and solar energy. It show a decrease of about 7% in 2011 compared to 2010 as wind and solar power intensity decreased. As shown in Figure 5, the year 2006 was less favourable to hydropower production, because it

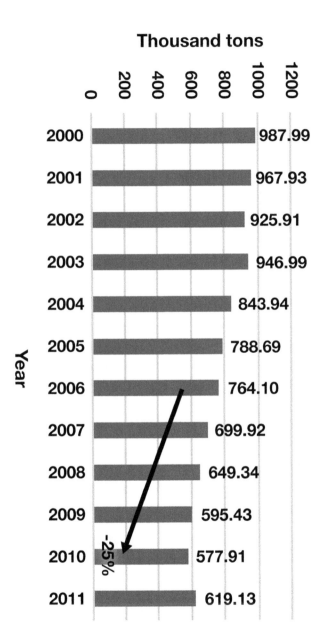

Figure 3. CO_2 emission intensity of energy in Romania between 2000 and 2011.

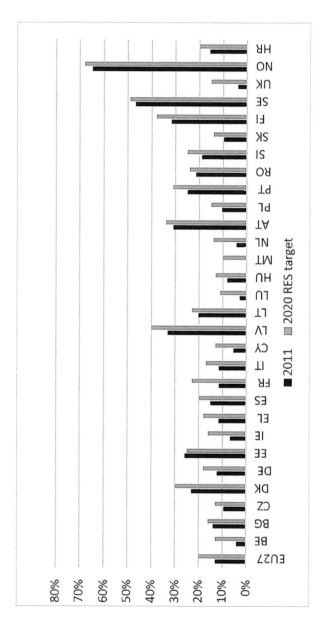

Figure 4. Member States progress towards 2020 targets in renewable energy, % [19].

was drier than 2005. The low rainfall in the coming years led to a decrease in the yearly energy production from renewable sources.

In terms of solar radiation in Romania, the monthly spread of the values on the Romanian territory reaches maximum values in June (1.49 kWh m^{-2} day^{-1}) and minimum values in February (0.34 kWh m^{-2} day^{-1}). Most solar-thermal systems are made with flat or vacuum-tube solar collectors, especially in areas with low solar radiation in Europe. In the national strategy on the use of renewable energy sources, the stated wind potential is 14,000 MW (installed capacity), which can provide a quantity of energy of approximately 23,000 GWh $year^{-1}$. These values are an estimate of the achievable potential. With respect to the energy potential of the biomass, the Romanian territory was divided into eight regions, accumulating in the year 2011 around 3.618 million tonnes, thus having the largest share of the total renewable energy resources in Romania. Water resources of developed inland rivers are valued at about 42 billion cubic metres per year, but because they are undeveloped, Romania can rely on having approximately 19 million m^3/year due to fluctuations in the flow of rivers. The synthesis of the achievable RES potential of Romania is analysed in Table 1 [20]. For "tonne" we have used the abbreviation "toe" throughout the entire paper.

2.2.2 THE TRANSPORTATION SECTOR

Rules and policies in energy and the environment highlight the considerable environmental impact of urban agglomerations and increase in the number of motor vehicles. According to the latest studies, urban traffic generates 40% of CO_2 emissions and 70% of other pollutant emissions [21,22,23,24,25].

At the EU level, transportation is responsible for about 28% of GHG emissions, and 84% of these are caused by road transport. The high level of emissions from road transport (CO_2, CO, NO_x, SO_2, NH_3, and volatile organic compounds, particles loaded with heavy metals, i.e., lead, cadmium, copper, chromium, nickel, selenium, and zinc) has a

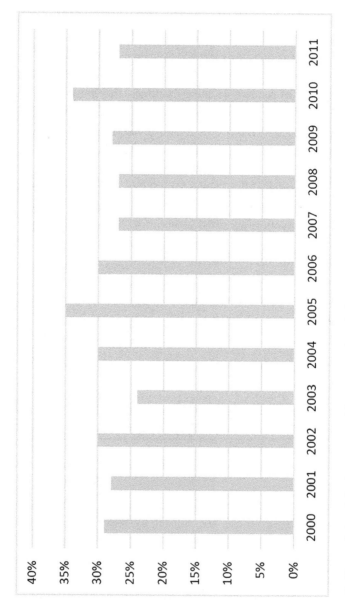

Figure 5. Share of electricity from renewable sources of total electric power [6].

Table 1. Potential of Romania for renewable energy sources.

Renewable Energy Source	Technical Potential [GWh]	Technical Potential [thousand toe]
Wind	409,731	35,231
Solar thermal	14,932	1284
Solar photovoltaic	161,929	13,923
Biodiesel	6084	523
Bioethanol	45,461	3909
Solid fuel	71,966	6188
Micro hydro	14,724	1266
Geothermal	279	24
Total	710,103	61,058

considerable effect on the environment, human health and on local and national sustainable development.

Eurostat highlights that, between 2011 and 2012, CO_2 emissions decreased in nearly all Member States, except Malta (+6.3%), the United Kingdom (+3.9%), Lithuania (+1.7%) and Germany (+0.9%). The most significant decreases were those in Belgium and Finland (both −11.8%), Sweden (−10.1%), Denmark (−9.4%), Cyprus (−8.5%), Bulgaria (−6.9%), Slovakia (−6.5%), the Czech Republic (−5.2%), Italy and Poland (both −5.1%).

In Romania, in 2010, there were 32,897 (thousand tons CO_2) going down in 2012 and reaching 30,758 (thousand tons of CO_2). Thus, there is an improvement of 6.5%, i.e., a decrease in CO_2 emissions of 2140 (thousand tons CO_2) [26]. This CO_2 decrease is due primarily to the legislative environment by introducing the environment tax for aggressive polluting cars and due to investments in educating people about the use of renewable resources at an industrial and domestic level (by total or partial funding of using these emerging technologies). EU reports that, in

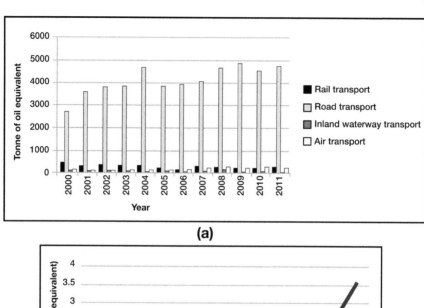

(a)

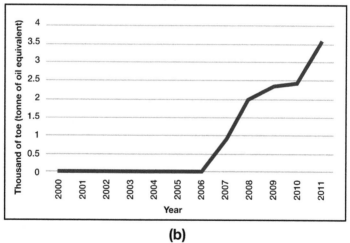

(b)

Figure 6. (a) Energy consumption by transport mode (national/modes of transport/1000 toe). (b) Biofuel consumption in the transport sector (%/national/1000 toe) [6].

terms of fuel consumption, the 2015 target is approximately equivalent to 5.6 L per 100 km of petrol or 4.9 L per 100 km of diesel. The 2021 target equates approximately to 4.1 of petrol or 3.6 L/100 of diesel litres per 100 km [20,26].

Transport systems used in Romania are freight and passenger transport. Within these systems, the following networks operate: road transport, rail transport (maritime and inland waterways), air transport, non-motorized and special (through pipes and electric air transport).

Energy consumption for every mode of transport in Romania is shown in Figure 6a [6]. Thus, road transport has the greatest share, around 89% of the total energy consumption in transportation, in 2011.

To improve the current situation, efforts are being made for reducing the number of and retiring the old, highly-polluting vehicles from the roads and for developing strategies to encourage the population to purchase hybrid cars.

As seen from the evolution presented in Figure 6, steps are being taken to support the use of bio-fuels and hybrid cars in transportation, which lead to a value of 3.55% in 2011 as shown in Figure 6b. The European Union Directive 2003/30/EC presents the objective of reaching a 5.75% share of renewable energy as a proportion of the total energy consumption of the transport sector by 2010 [27].

It should also be mentioned—even though it is not within the scope of this paper—it has been argued that the 2030s will bring about transport technologies (with low or zero emissions) that will be implemented on a large scale, and the following decade (the 2040s) will call on major and complex decisions to be made on energy technologies and on the structure of the EU economies.

2.2.3 THE WASTE SECTOR

Good waste management based on waste selective collection and recycling can decrease the carbon dioxide [28,29,30,31]. In Romania, the municipal solid waste (MSW) recycling rate reached its highest value in 2011, i.e., 7% of the total collected waste. The evolution of MSW collection and recycling are shown in Table 2. The quantity of MSW collected per capita, in the year 2011, in Romania was 239 kg, 22% less than in 2008. In Romania, waste is mostly stored in locations specifically designated for this operation, but which are not managed properly. However, some pilot experiences on

Table 2. Municipal waste collection and recycling rate [6].

MSW (tons)	2006	2007	2008	2009	2010	2011
Collected	6,334,491	6,187,943	6,558,342	6,264,778	5,325,808	4,553,300
Recycled	40,945	65,741	72,110	100,455	296,342	331,622
MSW recycling rate	0.65%	1.06%	1.10%	2%	6%	7%

composting, thermal treatment or co-combustion of some fraction of MSW as they are or pre-treated were developed or are in progress [32,33].

2.2.4 THE INDUSTRIAL SECTOR

The industrial sector is a major polluter in Romania, among the top three sectors that emit CO_2 into the atmosphere, contributing to environmental pollution. In Romania, according to public reports, in 2010, there were 491,805 active companies, significantly less than in 2008 (Figure 7a). Currently, the number of active companies is in a slight decline, which is the main cause of the reduction of waste generated by economic activity. In the current situation of economic instability, there is a downward evolution in the number of active companies as reported by the national statistics [34]. The first waste generator is mining, but the quantity has decreased since 2007 (Figure 7b).

2.3 RESULTS AND DISCUSSION

The above data may be useful in performing an assessment of the trends in the use of renewable resources in Europe and Romania. Thus, in the EU, considerable increase in the use of RES is expected, in order to mitigate the environmental impact of greenhouse gases, and in particular of carbon dioxide. Figure 8 shows the evolution of RES in the analysed sectors in

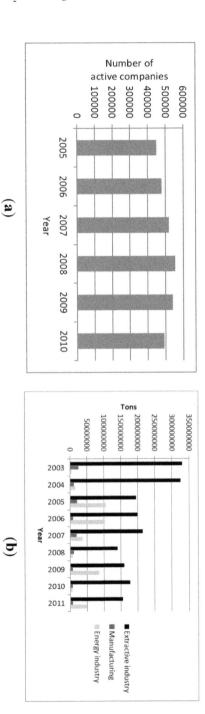

Figure 7. (a) Romanian active companies (b) Generated waste by economic activity (tons).

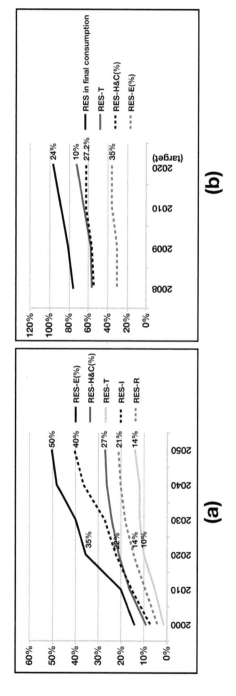

Figure 8. (a) Renewable energy resources (RES) indicators at the EU level. (b) RES indicators at the Romanian level.

accordance with EU strategies: renewable energy sources for electricity (RES-E), renewable energy sources for heating and cooling (RES-H&C), renewable energy sources for transport (RES-T), renewable energy sources for industry (RES-I) and renewable energy sources for residential (RES-R). RES-E is clearly the most important with an expected value of 50% until 2050 and an increase of 35% until 2020 [34,35]. Assessing Romania in terms of the use of RES and the alignment with the EU strategies and policies, the situation is shown in Figure 8b.

Analysed data related to GHG emissions proves that there will be a difference of at least 50 million tons of CO_2 equivalent annually, as difference between the target value laid down in the Kyoto Protocol and the total emissions in the commitment period of the 2020 convention, even considering the possible uncertainties related to inventories and projections of GHG emissions. The economic crisis of the recent years has reduced these levels even further [35]. The total GHG emissions of Romania in 1990 were 253.3 (million tons), being reduced by 52.1% in 2010, i.e., 121.4 (million tons).

According to the Kyoto Protocol, average 2008–2011 emissions in Romania were 53% lower than the base-year level, well below the Kyoto target of −8% for the period 2008–2012. LULUCF activities are expected to decrease net emissions by an annual amount equivalent to 1.1% of base-year level emissions.

In the second part of this section, Romania's carbon footprint is to be shown. Thus, CO_2 emissions in Romania highlights the environmental impacts associated with energy use in various sectors of activity listed at the beginning of this study. The total CO_2 emissions in Romania is divided by the total number of citizens and the thus the carbon footprint indicator is obtained; it represents virtually the total amount of greenhouse gases (expressed in CO_2) we produce per year through the burning of fossil fuels for heat or electricity we consume, as shown in Figure 9.

2.4 A SUMMARY OF FINDINGS

A number of considerations may be made on the basis of available data, starting from the situation of Romania's in comparison with that of the EU

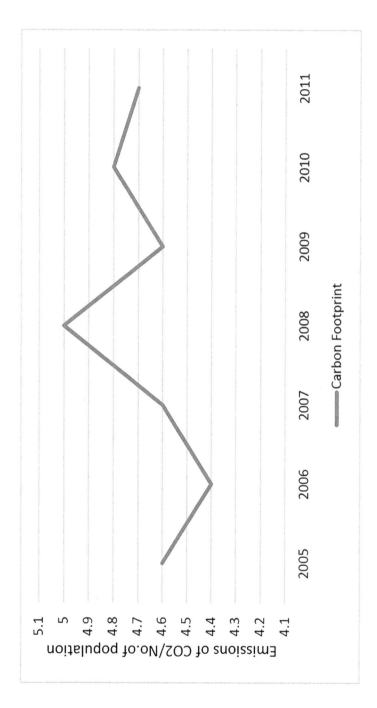

Figure 8. (a) Energy consumption by transport mode (national/modes of transport/1000 toe). (b) Biofuel consumption in the transport sector (%/national/1000 toe) [6].

and considering the participation in the Kyoto Protocol and the Covenant of Mayors-EU:

- Level of total GHG emissions is lower compared to 2008, but there was a small increase in 2011 (according to Figure 1a);
- The emissions decreased mainly in public electricity and heat production and road transport (according to Figure 1b and Figure 2) ;
- In terms of the CO_2 emissions in Romania, there are actions to monitor and reduce pollution in excessively polluting sectors (the first three sectors): electricity and heat production, manufacturing industry and transport. For example, Romania's energy sector emits 8% more CO_2, i.e., 48% (According to Figure 2). These actions and strategies mainly refer to the use of renewable sources (subsidized by the government) and to the reduction of polluting factors and their replacement with innovative elements;
- Due to the intensity of rainfall and climate change, the share of electricity from renewable sources of the total electric sector was 27% in 2011, i.e., a lower value than in 2010 when it was 34% (according to Figure 5);
- In terms of transport, both in Romania and the EU, road transport has the largest share of modes of transport. It consumes the greatest amount of energy, surpassing by far air, rail or inland waterway transport (according to Figure 6a and [6]);
- There are efforts to use bio-fuels and hybrid cars in transportation, reaching a value of 3.55% in 2011, which was a major breakthrough compared to 2006 (according to Figure 6b);
- The waste sector is in an unacceptable situation in comparison with the EU. The expected reduction of landfilling will positively affect the role of MSW sector in the balances of GHG thanks to the decrease of fugitive emissions of methane (according to Figure 7a,b);
- Romania is situated below the EU per capita average amount of MSW generation;

- The industrial sector shows a slight decrease in the number of active companies, and thus there was a slight decrease in the amount of waste generated.

Considering the above, Romania is making substantial efforts to reduce CO_2 emissions by adopting nationwide the necessary directions and strategies (for example, subsidies granted to producers of renewable energy, reducing national taxes for cars that do not pollute excessively, by purchasing vehicles for public transport that run on renewable energy, and others), thus aligning to European directions for the use of renewable energy and adoption of effective waste management policies.

2.5 CONCLUSIONS

The effectiveness of resources must be examined in the context of sustainable development. There is a strong connection between the quality of life and the way countries manage their natural available resources. Resource-efficient countries have a higher degree of innovation, productivity at lower cost and low impact on the environment, while providing multiple opportunities for improving emissions and sustainable life styles. Therefore, the efficient use of resources is based on a number of factors, such as redefining the way in which urban systems are globally understood, developing a common language for the assessment of sustainability indicators, reviewing the indicators which constitute the sustainability of cities and the financial support of the entities.

The main greenhouse gas produced by human activities is carbon dioxide (CO_2). It represents over 80% of the total emissions of greenhouse gases in EU Member States. The presentation of Romania's situation globally represented the starting point for future research: evaluation of CO_2 emissions for each mode of transport, the investigation of waste management technologies and their impact on Romania, the assessment of the types of industries and amounts of greenhouse gas emissions and the modelling of renewable resources at a national level.

REFERENCES

1. United Nations Conference on Environment and Development—UNCED. Agenda 21. Available online: http://www.un.org/esa/dsd/agenda21/ (accessed on 2 November 2014).
2. Häikiö, L. Institutionalization of Sustainable Development in Decision-Making and Everyday Life Practices: A Critical View on the Finnish Case. Sustainability 2014, 6, 5639–5654.
3. European Commission. Available online: http://ec.europa.eu/ (accessed on 10 May 2013).
4. McGlade, J.M. State of the Water Environment. In Proceedings of the Informal meeting of Ministers of Environment and Climate Change, Nicosia, Cyprus, 7–8 July 2012.
5. Varga, B.O. Electric vehicles, primary energy sources and CO2 emissions: Romanian case study. Energy 2013, 49, 61–70. [Google Scholar] [CrossRef]
6. National Institute of Statistics. National Statistical System. Available online: http://www.insse.ro/cms/en (accessed on 2 November 2014).
7. International Energy Agency. IEA Statistics. CO2 Emissions from Fuel Combustion, 2013 Edition. Available online: http://www.iea.org/publications/ freepublications/publication/co2emissionsfromfuelcombustionhighlights2013.pdf (accessed on 2 November 2014).
8. Regional Development Agency. Statistical communications. Available online: http://www.adrcentru.ro/ (accessed on 2 December 2014).
9. Covenant of Mayors. Sustainable energy action plans. Available online: http:// www.covenantofmayors.eu/ (accessed on 2 December 2014).
10. Duflou, J.R.; Sutherland, J.W.; Dornfeld, D.; Hermann, C.; Jeswiet, J.; Kara, S.; Hauschild, M.; Kellens, K. Towards energy and resource efficient manufacturing: A processes and systems approach. CIRP Ann.–Manuf. Technol. 2012, 61, 587–609.
11. Yuan, C.; Zhai, Q.; Dornfeld, D. A three dimensional system approach for environmentally sustainable manufacturing. CIRP Ann.–Manuf. Technol. 2012, 61, 39–42.
12. MoosaviRad, S.H.; Kara, S.; Hauschild, Z. Long term impacts of international outsourcing of manufacturing on sustainability. CIRP Ann.–Manuf. Technol. 2014, 63, 41–44.
13. Paska, J.; Surma, T. Electricity generation from renewable energy sources in Poland. Renew. Energy 2014, 71, 286–294.
14. Swider, D.J.; Beurskens, L.; Davidson, S.; Twidell, J.; Pyrko, J.; Pruggler, W.; Auer, H.; Skema, R. Conditions and costs for renewable electricity grid connection: Examples in Europe. Renew. Energy 2008, 33, 1832–1842.
15. Ostergaard, P.A. Reviewing optimization criteria for energy systems analyses of renewable energy integration. Energy 2009, 34, 1236–1245.

16. Lund, H.; Mathiesen, B.V. Energy system analysis of 100% renewable energy systems—The case of Denmark in years 2030 and 2050. Energy 2009, 34, 524–531.

17. Ban, M.; Perkovic, L.; Duic, N.; Penedo, R. Estimating the spatial distribution of high altitude wind energy potential in Southeast Europe. Energy 2013, 57, 24–29.

18. Gaigalis, V.; Markevicius, A.; Katinas, V.; Skema, R. Analysis of the renewable energy promotion in Lithuanian compliance with the European Union strategy and policy. Renew. Sust. Energ. Rev. 2014, 35, 422–435.

19. European Commission. Europe 2020 Indicators—Climate Change and Energy; The Commission to the European Parliament and the Council: Bruxelles, Belgium, 2014.

20. Dusmanescu, D.; Andrei, J.; Subic, J. Scenario for implementation of renewable energy sources in Romania. Procedia Econ. Financ. 2014, 8, 300–305.

21. Ryu, B.Y.; Jung, H.J.; Bae, S.H. Development of a corrected average speed model for calculating carbon dioxide emissions per link unit on urban roads. Transp. Res. Part D Transp. Environ. 2015, 34, 245–254.

22. Park, M.S.; Joo, S.J.; Park, S.U. Carbon dioxide concentration and flux in an urban residential area in Seoul, Korea. Adv. Atmos. Sci. 2014, 31, 1101–1112.

23. Rada, E.C. Sustainable city and urban air pollution. WIT Trans. Ecol. Environ. 2014, 191, 1369–1380.

24. Kheirbek, I.; Ito, K.; Neitzel, R.; Kim, J.; Johnson, S.; Ross, Z.; Eisl, H.; Matte, T. Spatial variation in environmental noise and air pollution in New York City. J. Urban Health 2014, 91, 415–431.

25. Istrate, I.A.; Oprea, T.; Rada, E.C.; Torretta, V. Noise and air pollution from urban traffic. WIT Trans. Ecol. Environ. 2014, 191, 1381–1389. [Google Scholar]

26. Eurostat Statistics. Available online: http://epp.eurostat.ec.europa.eu/ (accessed on 2 December 2014).

27. Eurostat Statistics. Available online: http://ec.europa.eu/energy/observatory/ trends_2030/doc/trends_to_2050_update_2013.pdf (accessed on 2 December 2014).

28. Fujii, M.; Fujita, T.; Ohnishi, S.; Yamaguchi, N.; Yong, G.; Park, H.S. Regional and temporal simulation of a smart recycling system for municipal organic solid wastes. J. Clean Prod. 2014, 78, 208–215.

29. Ionescu, G.; Stefani, P. Environmental assessment of waste transport and treatment: A case study. WIT Trans. Ecol. Environ. 2014, 180, 175–185.

30. Panepinto, D.; Genon, G. Carbon dioxide balance and cost analysis for different solid waste management scenarios. Waste Biomass Valoriz. 2012, 3, 249–257.

31. Rada, E.C.; Squazardo, L.; Ionescu, G.; Badea, A. Economic viability of SRF co-combustion in cement factory. UPB Sci. Bull. Serie D 2014, 76, 199–206.

32. Ghinea, C.; Petraru, M.; Bressers, H.T.A.; Gavrilescu, M. Environmental evaluation of waste management scenarios—Significance of the boundaries. J. Environ. Eng. Landsc. 2012, 20, 76–85.

33. Negoi, R.M.; Ragazzi, M.; Apostol, T.; Rada, E.C.; Marculescu, C. Bio-drying of Romanian Municipal Solid Waste: An analysis of its viability. UPB Sci. Bull. Serie C 2009, 71, 193–204.

34. Government of Romania—Ministry of Foreign Affairs. National Report 2013. Evaluation of the impact of reducing emissions of greenhouse gases on the Romanian economy. Available online: http://ec.europa.eu/europe2020/pdf/nd/nrp2012_romania_en.pdf (accessed on 2 December 2014).
35. Wadim, S.; Štepán, K.; Evgeny, L. Energy Economics and Policy of Renewable Energy Sources in the European Union. Int. J. Energy Econ. Policy 2013, 3, 333–340.

Barriers to the Deployment of Low Carbon Technologies: Case Study of Arun™ 160 Solar Concentrator for Industrial Process Heat

PLEASA SERIN ABRAHAM AND HARIPRIYA GUNDIMEDA

3.1 INTRODUCTION

The suggestion of IPCC (2007) to reduce carbon emissions by 50% - 85% by the year 2050 clearly shows the need of transition towards a low carbon economy [1]. It challenges an economic model which is heavily carbon dependent from the period of industrial revolution. The record of growth of GDP in the last 150 years shows its strong positive correlation with the increasing carbon emissions. Therefore delinking economic growth and fos- sil fuel use conflicts with existing pattern of investment and consumption and alternate models face the historical condition of "Carbon Lock-In" [2].

The carbon intensive economies and social systems show some inertia towards any kinds of policy which demands the diffusion of environment-superior technologies. The alternate technologies are unable to start up because the current path of fossil fuel based system is showing path dependent increasing returns to scale [3].

The highly evolved Techno Institutional Complexes (TIC) in developed (industrialized) economies prevents them from switching to environmentally feasible and economically efficient technologies. The relative stability of the standardized technological system is due to the irreversibility of the investments made by several generations in the infrastructure. Establishment of dominant design will lead to a shift occurring from product (Schumpeterian innovation) to process (Usherian) innovation. Incremental improvements in design, market driven R & D, specialization and development of core competency of the firm, management and organizational practices which nurture it will lead to standardization of the technology. Also the capital investments go to the area where production costs and uncertainty are low and risk-averse lending practices will fund the standardized technologies. The professions, disciplines etc. based on this technological system preserve the technology along with unions and industry organizations which have the same interests of the oligopolistic firms. The state and its policies ascertain the existence of such system which ultimately leads to the standardization of the system. All this postpones the eventual obsolescence and substitution [2].

Economic theory deals with technological obsolescence but not with the system obsolescence. Because of the inflexibilities new innovators in the area of clean technology faces excess inertia since they have to compete with the standardized models. This results in the persistence of multifaceted barriers in this field.

The bottom up engineering approaches have come up with "non-conventional" energy sources powered equipment which in long term can challenge the fossil fuel based system if current incentive structure in the market. This can be showed by the recent developments in the solar, wind energy utilization and development of hydrogen cells. But economic modeling contrary to engineering one is top-down and it assumes that at present the economy is functioning efficiently in equilibrium and any reduction in carbon emissions will only happen at an expense of economic activities in the economy.

There is a significant lag between dominant technological practices prevailing and technically feasible technologies. Many of the technologies which succeed in laboratory cannot do so in the market. Innovations which speedily cross the valley of death will have an easy deployment space. Till the demonstration succeeds the "era of ferment" prevails and this era of turbulence or disturbance is due to uncertainty about the performance of the technology. A state funded R & D or investment by venture capitalist helps it to march from laboratory to market [4] . Once the confidence is instilled market deployment happens and it has to complete with standardized technologies in the market. Figure 1 shows the stages of journey of the technology from laboratory to market. After the development, it has to be financed by agencies like government, angel investors etc. who foresee the potential of the technology. Venture capitalists who adopt the technology help in the reduction of risks and removal of barriers. The agency which takes up the technology in its early stage has to pass through a period of negative cash flow and possibility of failure in market dissemination and success.

Technologies developed principally to mitigate carbon emissions face additional challenges in the "Valley of Death" because technologies to reduce them generally do not have existing markets to produce capital to "pick them up" on the other side [5]. If the technology survives the valley of death phase, it is actually deployed in market where it has to compete with the conventional technologies. Like product differentiation, technology differentiation is not a key market driver and therefore it has to face a lot of barriers in an unlevel playing field (Brown, 2007). This study is on Arun™ 160 technology (Only paraboloid concentrator technology developed in India till date for industrial process heat applications) which didn't get choked off in the "valley of death" and the barriers it faces in its market deployment despite of the economic and environmental advantages it offers.

3.2 ARUN SOLAR CONCENTRATOR
FOR INDUSTRIAL PROCESS HEAT SYSTEM

The industrial sector in India is large and diverse, encompassing some 150,000 manufacturing firms that employ more than 9 million people [7] India consumes 2,722,000 bbl of oil per day and 40% of this oil

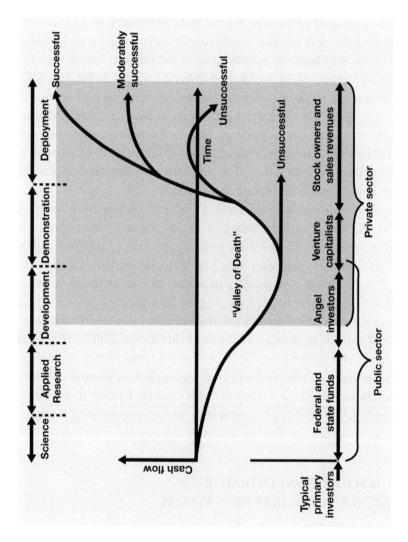

Figure 1. Innovation and market. Murphy and Edwards (2003) [5] cited by Brown (2007) [6] .

consumption is in the industry and in that almost 60% - 70% of industrial use is in thermal form and 70% of that is demanding below 250 degree Celsius [8]. Most of the industries like diary, meat and food preservation industries use fall into the requirement of this energy. A large number of these applications are for small scale industries for process heat, cooling, water desalination.

Arun™ 160 is a supply side technology that is an indigenous version of Solar Concentrator technology for Industrial Process Heat (IPH). It is developed by IIT Bombay with the support from MNES (Ministry of New Energy Sources) of Indian government. The technology which is patented and marketed by Clique Energy solutions; an Engineering Consultancy firm has the potential to undertake structural transition in many industries by capping carbon emissions and by reducing oil bills.

The first Solar Concentrator dish for Industrial Process Heat system from Arun family was installed in Latur, Maharashtra under R & D project of IIT Bombay with Clique Developments Private Limited. It was sponsored by MNES with Mahanadi Diary contributing 50% of the Solar System cost and Clique bearing part of the design cost. They have successfully installed and commissioned Arun™ 160 dish that can generate process heat at about 200°C, store it and supply it at desired process temperature any time of the day or night. The thermal energy delivery to the plant for milk pasteurization process is in the range of 60 - 80 kW and the average energy delivered is in the range of 1900 - 2200 MJ/day on clear days. The average system efficiency, including the thermal losses from piping, fittings etc., based on normal beam solar radiation incident on the aperture plane of the collector is 51.4% on clear sunny days. The solar concentrator Arun™ 160 is able to save about 60 - 65 l of furnace oil on a clear day. The annual savings of furnace oil is about 17,500 l with Arun operating for about 275 - 290 days/yr [9] .

The annual working hours of Arun™ 160 is 3200 - 3350 h/yr. The annual fuel savings include 16,000 - 24,000 l/yr and electricity savings include 140 - 180 MW/yr and annual reduction of carbon dioxide emissions are 42 - 200 t [10] .

Till date this is India's solar concentrator for Industrial Process Heat with largest aperture area and highest annual heat generation capacity, highest operating temperatures and highest stagnation temperatures and capacity of day-long energy storage and integration with a wide range of applications. The Arun™ Solar Concentrator System can be used in "ADD ON" mode and can be retrofitted to the existing boiler or heater system in the industry. This maintains the "continuity principle" of a new technology which reduces cost of replacement and brings down the psychological costs [11].

It can be used for providing process heat for a wide range of industries and chemical processing plants using boilers or heaters, textile mills, sugar mills, vegetable oil mills, agro and food processing industries, timber industry, milk processing, drying of horticultural, food and fruits products, drying of chemicals as well as units using vapor absorption refrigeration for space cooling. It is also suitable for hotels and hospitals for providing hot water, steam and cooling. This shows that it is not customized but same technology can revolutionize the energy demand in industrial field by replacing the existent technology and reducing emissions.

The Clique energy solutions group has developed the design of Arun Solar Concentrator system to cater to the growing thermal needs of industry by harnessing clean solar energy using commercially viable technology. The prototype model was successfully tested in May 2003 and based on the experience a commercial unit with a reflector area of 160 sq meters has now been installed for use in the Mahan and dairy in Maharashtra. This was a successful demonstration project which led to the take off technology in the market. After that the technology was identified to be mature enough to get into markets and from 2006 onwards Clique actively started marketing the technology.

However the technology is facing a lot of barriers that it got only few customers in spite of its wide range of scope. From 2006-2010 they had only sold nine dishes to four customers. From our interaction with the marketing department of this technology we came to know that almost 400 firms, both private and government firms have approached them to know more about the technology but were reluctant to take up the technology. Some of them were at various stages of discussion due to several hindrances they got into while taking this decision. In this case study, a sample of potential customers who dropped the idea of taking up Arun™ 160 has been approached to understand the barriers faced by this technology.

3.3 CLASSIFICATION OF BARRIERS

In this paper we have defined "Technology Barriers as the obstacles to private investment" and thus considered as a pull factor that inhibits private investment in this technology [12]. Barriers are divided into three according to this typology [12]. The barriers have been classified into Micro barriers, Meso barriers and Macro barriers.

3.3.1 MICRO BARRIERS

These are technology specific barriers, which create obstacles that are unique to a particular project. The micro barriers can specifically be in terms of project design, which affects the feasibility of the project. By changing the features of the project, modifying design, improving energy saving features, giving confidence through proper consultation etc such barriers can be reduced or removed. Cost of the technology falls in this category as it is varies from one technology to other especially upfront costs.

3.3.2 MESO BARRIERS

These relate to the organization or firm level barriers such as lack of incentive for energy policy, absence of organization environmental policy, strict budget management policies etc. This relates to organization affiliated with the project and happens in the implementation stage. These can be tackled by split incentives, re training of energy department staff etc.

3.3.3 MACRO BARRIERS

These can be the barriers that exist due to the state policies; market related and can be even civil society related. For project designs and organization, they are external barriers and firms cannot influence them unless they have the power to influence politics, market or culture. Barriers related to state are visible in government policies, laws, ministry declaration, subsidy allocation etc while market related barriers include reluctance of private banks to finance new technology, hidden information etc. Barriers relating to civil society include the behavior and attitude of NGOs, academic institutions etc.Table 1 shows the typology of barrier used in the study and gives an idea why it could be a possible barrier that can inhibit the private investor in adoption of the technology.

3.4 METHODOLOGY

A qualitative study was done to observe the barrier and their intensity. To investigate the views of different stakeholders on barriers hindering the introduction and implementation of Arun 160 we have conducted a two round study. The study has been carried out in three stages:

1. Identifying barriers;
2. Constructing the questionnaires for firms and experts and collecting data;
3. Comparing the results of both the rounds.

The study was carried out in two rounds: one with firms and another with experts. The questions were based on the possible applications of technology, barriers that led to rejection of the technology. Barriers were rated into quantitative scale according to the intensity of the barrier as per the response. The mean weight of each one is categorized and compared to order it according to the intensities. In the second round expert's opinion were taken into consideration. They include experts who were in the project right from the stage of development and marketing. Their ratings were weighed and once again mean weight was calculated. The rank score is given according to the rank secured by the barrier. Then rank scores secured by each barrier in both the rounds are added to get the barrier intensity.

Nine firms from different manufacturing areas like engineering, dairy, automobiles, hotel, distillery, construction etc. who have not implemented

Table 1. Barriers to the adoption of technology.

No	Barriers	Description
B1	Micro barriers	
B1.1	Space constraint	Huge aperture area of dish creates space crunch for firms.
B1.2	Geographical reasons	Solar devices may not be equally efficient in all areas.
B1.3	High upfront costs	High initial capital costs as compared to conventional technologies.
B1.4	Low scale of technology	Low production volume of energy compared to the needs of firm.
B1.5	Skepticism on performance efficiency	Psychological costs are high when there is lack of network externalities and positive feedbacks.
B2	Meso barriers	
B2.1	High transaction costs	Costs of identifying, assessing and observing them become costly.
B2.2	Cost of staff replacement and training	Costs on training and bringing up a new technical labor force.
B2.3	Management norms on capital budget	Low priority given to investment in unproven technologies.
B2.4	Technical skills and staff awareness	Lack of awareness on renewable and energy efficient technologies.
B2.5	No incentive for energy savings	Lack of incentive within the firm for energy cost reduction.
B2.6	Lack of energy and environmental policy in firm	Absence of energy and environmental policies which help to look for alternate technologies.
B3	Macro Barriers	
B3.1	Credit and soft loan availability	Banks discourage credit and soft loans given to unproven technologies.
B3.2	Business market uncertainty	Market attitude towards new technologies when standard technologies are available.
B3.3	Lack of clarity on carbon credits	Uncertainty and tiring procedures on carbon credits create confusion.
B3.4	Uncertainty about subsidy	Policy uncertainty on subsidy given to this technology.

the technology have responded and participated in the first round of study. We have contacted the 400 firms who were in discussion with the Clique Energy solutions who were marketing the technology. All these industries have varied applications of IPH (Industrial Process Heat). In this paper an effort has been made to understand why these firms did not implement the technology. From the questionnaires we have sent we got back 26 filled ones and among them only nine were valid. The questionnaire asks specifically the kind of barrier they faced which led to the rejection of this technology and analyses the vulnerability of firm to a particular barrier.

In the second round three experts have participated and have rated the barriers according to their intensity. The three experts who have worked in the research, development and marketing of the technology have filled the questionnaire to complete the second round.

Table 2 shows will give an idea about the profile of firms and experts who have participated in the survey. There were two stakeholder groups one being the firms themselves and other experts.

3.5 RESULTS AND DISCUSSIONS FROM BOTH THE ROUNDS

Among the micro barriers high upfront costs is the most serious barrier as they have to compete with the standardized technologies in market (Figure 2). The cost of one dish is sixty lakh rupees (98.46 USD) and this makes the payback period very high say; three to five years. The skepticism about low IRR (Internal Rate of Return) lowers the prospect of investment. The 12% subsidy of cost per dish offered by MNES could not reduce the intensity of the barrier. The thing to be noted is that in both the rounds this became the most intense barrier. The space constraint gets a weightage to be the second most intense barrier as the huge aperture area of the dish keeps them reluctant to invest in this technology. Since most of the industries are facing the problem of space crunch, they walk out from investment in this technology.

The low scale of technology and skepticism on performance efficiency got equal weightage. Most of the firms feel that the production volume of their technology is very low compared to their needs. The lack of network externalities and positive feedbacks makes the psychological costs to transition very high. The solar technologies need not be effective in all geographical regions as the solar insolation can be different and thus

Table 2. Representative profiles of the valid questionnaires.

Stakeholder groups	Number of valid responses	Nature of respondents
Enterprises	9	Representatives of firms who were potential customers but rejected the technology at different phases of discussion.
Experts	3	Researchers and sales managers who have worked in the development and promotion of technology.
Total	13	

geographical constraint though not relevant occupies a position but only in second round.

Micro barriers are technology specific barriers and unless and until this is removed or reduced the market cannot pick up. More attention has to be given to the improvements of specific features of technology which makes private investor more confident about the technology. Of course falling costs occurs when technology passes the "era of ferment". More research has to put in increasing the production volume and performance efficiency. Also improvements are to be made in the model so that it can be accepted by the companies.

Among the meso barriers, high transaction costs and strict management norms on capital budget gets equal weightage to become most intense barriers (Figure 3). The high transaction costs involve the information costs. The management instructions on capital budget only used to favor proven technologies and so there is no incentive to undertake a risky investment. Most of the firms think that the cost of replacement of staff and training is important. However interestingly, the other three barriers got weightage only in second round. While the firm believed that lack of awareness on renewable and energy efficient technologies, lack of incentive within the firm for energy cost reduction and absence of energy and environmental policies are not serious barriers the experts thinks the opposite. In the first round it was found that the firms had strong energy audit system and individuals are accountable for the energy costs so lack of incentive for energy savings and lack of policy on it doesn't make a barrier. But in the experts round, they got weightage.

The meso barriers show the inhibitions and fear the organization have towards change in the existing structure. More open policy to promote audit of energy and environmental costs has to be made from government to make firms accountable towards social costs. More information should be provided about green technologies by labeling and endorsing them from authentic sources. Proper marketing efforts can bring down these costs and proper monitoring of the technology and collection of positive feedbacks from the firms who have adopted the technology will bring down transaction costs and builds up network externalities. This will help to instill confidence and in long run will help the technology to get popular.

Among the macro barriers, uncertainty about the subsidy support and the long and tiring processes makes the most important barrier (Figure 4). This makes an important barrier from the side of government or ministry. Policy environment should be favorable that a speedy and certain way to subsidize the technology should be there.

The credit barrier becomes the major issue. Lack of soft loans for investment like this make the firms reluctant to invest in new technologies. It is high time for our banks to change their tastes from standardized conventional technologies to new energy resources. The offers on Carbon Emission Reduction (CER) credits are even more confusing and in the international level also there is a lot of uncertainty about the institution of Clean Development Mechanism and carbon credits. This forms a major institutional barrier and since Arun™ 160 technology has proved itself to be a carbon free technology, CER credits or other incentives which promote such technologies has to be given. The standard energy solutions which receive perverse subsidies make new technologies uncompetitive in market. This makes the business environment unfavorable towards new technologies.

There should be proper valuation of social costs and social benefits caused by technology to give it a level playing field in the market.

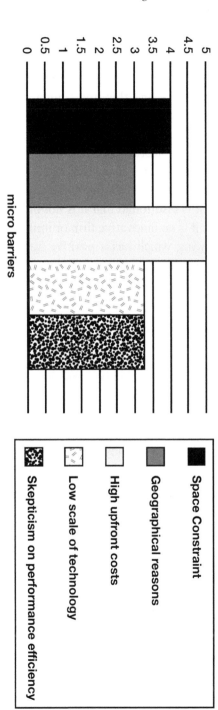

Figure 2. Micro barriers.

While discussing about the barriers it is also worth noting down the merits of this technology as pointed by some firms in our study. The huge uncertainty in energy costs have forced most of the firms to take up a second best alternative and this gives huge opportunities to renewable and unexplored technologies like Arun 160. Major automobile firm gives testimonial about the efficient design of solar concentrator and considers the foundation of project as perfect and systematic. The meticulous research on reduction of energy costs by the firms always brings down the information costs which will help in experimenting with new technologies. From the study it is also found that it is not only the size of a firm which decides whether it is an innovative firm or imitative firm. The firms which has a research wing which has a positive approach towards innovation takes more risk by taking up "out of the box" technologies despite of the financial and managerial obstacles they had to face.

3.6 LIMITATIONS OF THE STUDY

This is a highly perceptive study form the point of view of firms and experts. The size of the sample is not enough to make generalizations about the scope of the technology and many times the questionnaires received back were partially filled which made the discussion on various aspects of the industry missing in this study. There was difficulty in assembling the data on barriers in one single framework. A small degree of skepticism is there on the conversion of qualitative data into values. Also this study cannot make general conclusions on the barriers of other renewable energy technologies as this is technology specific.

3.7 CONCLUSIONS

The paper shows the intensity of barriers faced by an upcoming indigenous technology which has huge potential to cut the industrial carbon emissions by a large extent. Keeping in mind all the limitations of the study, some findings are worth noting down. Different approaches are required to

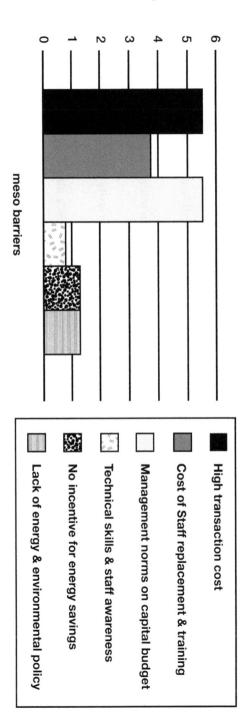

Figure 3. Meso barriers.

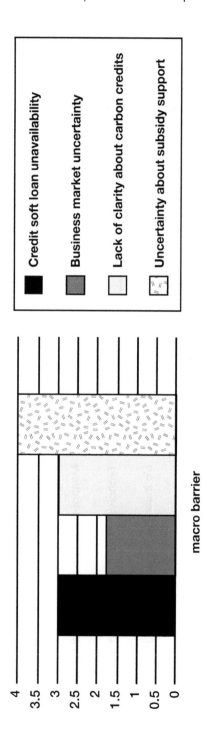

Figure 4. Macro barriers.

address each different barrier and without analyzing the nature and intensity of barrier policy decisions cannot be made.

The strategic niche management is very important because the "incubator technology" should be protected till the 'take off' stage [13]. The continuity approach is very much essential to lower down the financial and psychological costs of change and this particular technology scores there.

Funding R & D in developing energy efficient and new system design is a proactive idea, so that innovator is not hesitant even if Intellectual Property Rights are weak. Patents granted for innovations should be for an optimal time so that the diffusion process is not disrupted. Assistance to R and D will encourage firm level and individual level attempts towards building up new competitive and diverse models.

The transition management should have the strategic and long-term vision of the development of a technology from "niche to landscape". For that there should be an interaction between market and government. Government should offer place to market players by offering them "experimentation space". Government, market and society have to be partners in the process of setting policy proposals, creating opportunities and undertaking transition experiments [14]. The government becomes the facilitator and endorser by building networks and coalitions between actors and experimentation should promote diversity of technological options. When it comes to business stake holder, they should need clarity from government on future policy, long-term agenda on technology, technological development and transfer of technology. That is a removal of uncertainties to tailor its own business policy [15].

The climate paradox prevails because of the hidden information problem and the proper dissemination of information on energy efficient products should be there. Proper arrangements should be there to solve the problems of information asymmetry at the producer level and consumer level. The policy level approach to spread awareness about energy efficiency and potential of RETs is very important as it reduces the transactions costs in the economy.

The study shows the challenges of a carbon-saving technology in a country like India. The problem of carbon lock in is global and India is no exception. The multidimensional barriers faced by alternate technologies show that different types of approaches should be adopted to promote environment superior technologies. The consciousness has to rise from government level, market level and civil society level. And since anything related to breaking "carbon lock in" has a lot to do with technological innovations and improvement; so existence of a sustainable innovation regime is required [16]. After all tackling global warming will be the greatest technological project humans have to develop.

REFERENCES

1. IPCC (2007) Assessment Reports, AR4. http://www.ipcc.ch/
2. Harris, J. (2008) Ecological Macro Economics: Consumption, Investment and Climate Change. GDAE Working Paper No. 08-02 Ecological Macroeconomics.
3. Unruh, G.C. (2000) Understanding Carbon Lock-In. Energy Policy, 28, 817-830. http://dx.doi.org/10.1016/S0301-4215(00)00070-7
4. Arthur, B.W. (1989) Competing Technologies, Increasing Returns and Lock-In by Historical Events. The Economic Journal, 99, 116-131. http://dx.doi.org/10.2307/2234208
5. Murphy, L.M. and Edwards, P.L. (2003) Bridging the Valley of Death: Transitioning from Public to Private Sector Financing. National Renewable Energy Laboratory, Colorado.
6. Brown, A., Chandler, J., Lapsa, M.V. and Sovacool, B.K. (2007) Carbon Lock-In Barriers to Deploying Climate Change Mitigation Technologies. Oak Ridge National Laboratory. http://web.ornl.gov/sci/eere/PDFs/ORNLTM-2007-124_rev200801.pdf
7. Government of India (2010) Annual Survey of Industries 2005-06. Ministry of Statistics and Programme Implementation.
8. Key Issues for Renewable Heat in Europe (K4RES-H) (2006) Solar Industrial process Heat. European Solar Thermal Industry Federation, WP3 Task 3.5.
9. Kedare, S.B. (2008) Performance of ARUNTM 160 Concentrating Solar Collector Installed at Latur for Milk Pasteuri- sation. SESI Journal, 18, 1-9.
10. Bhave, A.G. (2012) Industrial Process Heat Applications of Solar Energy. International Journal of Modern Engineering Research (IJMER), 2, 3800-3802.
11. Unruh, G.C. (2002) Escaping Carbon Lock-In. Energy Policy, 30, 317-325. http://dx.doi.org/10.1016/S0301-4215(01)00098-2
12. Reddy, S.B., Assenza, G.B., Assenza, D. and Hasselmann, F. (2009) Energy Efficiency and Climate Change: Conserving Power for a Sustainable Future. Sage Publications, New Delhi.
13. Unruh, G.C. and Carrillo-Hermosilla, J. (2006) Globalizing Carbon Lock-In. Energy Policy, 34, 1185-1197. http://dx.doi.org/10.1016/j.enpol.2004.10.013

14. Foxon, T.J. (2007) Technological Lock-In and the Role of Innovation. In: Atkinson, G., Dietz, S. and Neumayer, E., Eds., Handbook of Sustainable Development, Edward Elgar, 140-152.

15. Foxon, T. and Pearson, P. (2008) Overcoming Barriers to Innovation and Diffusion of Cleaner Technologies: Some Features of a Sustainable Innovation Policy Regime. Journal of Cleaner Production, 16, S148-S161. http://dx.doi.org/10.1016/j.jclepro.2007.10.011

16. Neuhoff, K. (2008) Tackling Carbon: How to Price Carbon for Climate Policy. University of Cambridge, Cambridge. http://www.eprg.group.cam.ac.uk/wp-content/uploads/2009/03/tackling-carbon_final_3009082.pdf

PART III

OTHER GREENHOUSE GASES

CHAPTER 4

Atmospheric Ozone and Methane in a Changing Climate

IVAR S. A. ISAKSEN, TERJE K. BERNTSEN, STIG B. DALSØREN,
KOSTAS ELEFTHERATOS, YVAN ORSOLINI, BJØRG ROGNERUD,
FRODE STORDAL, OLE AMUND SØVDE,
CHRISTOS ZEREFOS, AND CHRIS D. HOLMES

4.1 INTRODUCTION

The chemically active climate gases ozone (O_3) and methane (CH_4) respond to variabilty in the current climate and will be affected by future climate change ([1,2]). O_3 and CH_4 will in addition to being influenced by atmospheric chemistry respond on changes in solar radiation, atmospheric temperature and dynamics, and are expected to play an important role for processes determining the interaction between the biosphere and the atmosphere ([2,3,4]). CH_4 chemistry is also affecting climate through its impact on ozone ([5]). Emissions of air pollutants and their precursors, which determine regional air quality by perturbing ozone and methane has also the possibility to alter climate. Climate changes will affect chemical processes in the atmosphere through transport on local and large scales and through removal and formation of pollutants ([6,7]).

Ozone is a secondary compound formed and partly broken down in the atmosphere, but also deposited at the Earth's surface, where it has a negative impact ([8,9]). High surface ozone levels harm humans through the respiratory system, and damage plant growth. Important changes in climate-chemistry interactions involving ozone occur via emission changes in ozone precursors and surface deposition of ozone due to changes in surface dryness ([1,10]). Furthermore, ozone is affecting the stratosphere through impact on dynamical chemical processes ([11,12,13]), and thereby on tropospheric gaseous distribution through modification of solar input ([14]). Enhanced oxidation is due to higher temperatures in synoptic high pressure systems with more sunlight favoring ozone production [15]. In addition, enhanced CH_4 emission from permafrost thawing in the Arctic and from other sources like wetlands and mining will enhance global methane and thereby ozone.

The chemical distribution, the oxidation potential and climate will be affected by O_3 and CH_4 perturbations ([4,16 17]). Perturbation of climate from O_3 and CH_4 changes will take place in the troposphere and in the stratosphere on local and global scales ([6]). Additional effects of ozone include reduced CO_2 uptake by plants ([9 18]).

Large methane changes will affect climate. For instance CH_4 seems to have been involved in a sudden warming that took place at the send of the Younger Dryas cold period with release from wetlands as a possible source ([19]). A potential important CH_4 source during permafrost thawing is the direct release from sub-sea deposits at the Siberian Shelf ([20,21,22]). Changes in CH_4 loss through the reaction with OH could also affect methane.

Emission of NOx from ships is estimated to increase OH in the background atmosphere and lead to enhanced loss of methane ([23,24]). Enhanced release of CH_4 and CO are of particular importance, since the reactions with these two compounds are major loss processes for OH, giving reductions in OH, thereby leading to further increases in CH_4.

In this article we will review studies of climate-chemistry interactions affecting current and future distributions of atmospheric O_3 and CH_4. Included in the review are one way studies of how the transport sector affects climate. We include studies by other groups and by the participating groups. In Section 2—"Atmospheric chemistry affecting ozone and methane in the atmosphere", some key chemical reactions (2.1) for O_3 and CH_4 in the troposphere and stratosphere are given, along with pollutant emission and their impact on ozone, methane and radiative forcing (RF) (2.2) and the impact of increased temperatures on tropospheric ozone levels (2.3). In Section 3—"Studies of climate-chemistry interactions"; IPCC activities (3.1), biosphere-atmosphere couplings (3.2), impact of methane emission from Arctic thawing (3.3), the impact of transport on the late winter ozone values in the Arctic in 2011 (3.4), and stratospheric ozone variability due to changes in dynamics (3.5) are presented. Section 4—"Conclusions" summarizes the findings.

4.2 PROCESSES AFFECTING OZONE AND METHANE

4.2.1 IMPORTANT OZONE AND METHANE CHEMISTRY

Shortwave solar radiation is essential for ozone production. Nitrogen oxides (NO, NO_2) and free hydrogen radicals (H, OH, HO_2) are involved in the ozone formation processes in sunlit regions (Equations (1)–(5)):

$$NO + HO_2 \rightarrow NO_2 + OH \qquad\qquad (1)$$
$$OH + CO \rightarrow H + CO_2 \qquad\qquad (2)$$
$$H + O_2 + M \rightarrow HO_2 + M \qquad\qquad (3)$$
$$NO_2 + h\nu \rightarrow NO + O \qquad\qquad (4)$$
$$O + O_2 + M \rightarrow O_3 + M \qquad\qquad (5)$$

Here, $h\nu$ represents the energy of a solar photon and M is a third body (usually O_2 or N_2).

A main ozone loss reaction in the troposphere, in addition to surface deposition, is the reaction with the hydrogen peroxy radical (Equation (6)):

$$O_3 + HO_2 \rightarrow 2O_2 + OH \qquad\qquad (6)$$

This reaction also represents a key process for OH production in the background atmosphere.

In the middle and upper stratosphere, ozone is produced through solar dissociation of O_2 (Equation (7)) followed by recombination of atomic oxygen with an O_2 molecule (Equation (8)):

$$O_2 + hv \rightarrow O + O \tag{7}$$
$$O + O_2 + M \rightarrow O_3 + M \tag{8}$$

In the stratosphere, ozone is lost mainly through reactions involving NOx and halogen-containing compounds (chlorine and bromine). Key reactions are of the type (Equations (9)–(11)):

$$O_3 + hv \rightarrow O + O_2 \tag{9}$$
$$O + OX \rightarrow O_2 + X \tag{10}$$
$$X + O_3 \rightarrow XO + O_2 \tag{11}$$

X could either be an OH, NO, Cl or Br radical.

Tropospheric O_3 can be lost through photolysis (Equation (9)) in the presence of water vapor (Equation (12)):

$$O(1D) + H_2O \rightarrow OH + OH \tag{12}$$

where $O(1D)$ is an excited state atomic oxygen atom. Besides the loss to hydrogen peroxy radical in Equation (6), another important O_3 loss in the troposphere is the reaction with hydroxyl (Equation (13)):

$$O_3 + OH \rightarrow O_2 + HO_2 \tag{13}$$

Equations (1), (9) and (12), with NO, O_3 and H_2O, respectively, represent key primary source of OH in the troposphere and lower stratosphere, while Equation (13) represents a loss reaction in the free troposphere. The main loss reaction for OH in the free troposphere is Equation (2) with CO.

Another important loss reaction for OH in the atmosphere is the reaction with methane (Equation (14)):

$$CH_4 + OH \rightarrow CH_3 + H_2O \tag{14}$$

CH_4 is a primary compound, emitted by different natural and anthropogenic sources (wetlands, rice production, livestock, mining, oil and gas production and landfills). The reaction with OH is the key loss reaction for atmospheric methane.

Through the reaction OH methane has a lifetime of the order of 8 years in the lower atmosphere (troposphere and stratosphere). CH_3 is oxidized rapidly in the atmosphere to yield O_3 (in the presence of NOx) and CO ([2,25]). Since CO is formed in the oxidation chain it will give further loss of OH. A minor fraction of methane will be removed through surface deposition (5% or less). Since the reaction with OH represents a main loss for CH_4 as well as an additional sink for OH, it affects the CH_4 lifetime. This process defines a feedback factor, which under current conditions is in the range 1.3 to 1.5 ([26,27,28]). The feedback factor could increase in the future if methane emissions increase. It has been shown that the increase in O_3 is non-linear with relatively higher impact on ozone for high CH_4 emissions ([20]).

4.2.2 POLLUTANT EMISSIONS AND THEIR IMPACT ON OZONE, METHANE AND RADIATIVE FORCING (RF)

There could be significant changes in the production of NOx from lightning as a result of changes in climate. Mickley et al. [29] considered earlier observations of ozone in connection with estimates of NOx emissions from lightning and impact on ozone and RF, and they concluded that the uncertainties in ozone production from lightning NOx emissions was larger than previously estimated.

Fiore et al. [6] looked at the impact of changes in lightning activity on the NOx distribution and furthermore on the ozone distribution. They found that emissions of air pollutants affect climate and that climate has an impact on chemical processes and on dynamical processes transporting pollutants from one region to a neighboring region. It was found that reducing the precursor CH_4 would slow near-term warming by decreasing both CH_4 and tropospheric O_3. There was significant uncertainty with the net climate forcing from anthropogenic nitrogen oxides (NOx) emissions, which increased tropospheric O_3 (warming) but also decreased CH_4 (cooling). Anthropogenic emissions of carbon monoxide (CO) and non-methane volatile organic compounds (NMVOC) were found to increase both O_3 and CH_4 because these compounds are ozone precursors, and increased methane since OH is reduced. A better understanding of how air pollution control influences climate is needed [30]. Comparisons with earlier studies were made. They also reviewed studies of the implications of projected changes in methane and ozone precursors for climate change and hemispheric-to-continental scale air quality.

Brasseur et al. [31] studied the impact of climate change on the future chemical composition of the atmosphere over the period 2000 to 2100 with the MOZART-2 model using meteorological fields provided by the ECAM5/Max Planck Ocean Institute Model. The study suggests that the impact on ozone of climate change is negative in large part of the troposphere as a result of enhanced destruction due to high levels of water vapor. The magnitude of the impact from climate change is smaller than the positive impact from emission changes. They also found that NOx levels from lightning production are increasing substantially.

Ozone levels are enhanced substantially when methane increases. Figure 1 shows the estimated ozone levels in the troposphere and UTLS (upper troposphere/lower stratosphere) region from Brasseur et al. [31] when the contribution from methane is included and not included.

Studies of the impact of precursor emissions on O_3 and OH in different regions, and from different transport sub-sectors have been performed for compounds like NOx, CO, CH_4 and NMHC, ([2,7,24,32,33,34,35,36,37,38,39]). In particular the effects from the transport sub-sectors aircraft and ship have been studied extensively ([2,6,7,24,32,34]).

Large scale ship emissions (remote areas) and aircraft (altitude range 8 km to 12 km) emissions occur in regions with moderate emission from other sources, while land based (road) emissions occur in areas where emission from other sources often are large. Estimates of the transport subsectors impact on O_3, CH_4 and on RF from the transport sectors show large differences in individual studies.

To illustrate the differences in impact from the transport subsectors we present the results from the study of [7]. They report results from 14 global chemistry transport models (CTMs). Regional distribution of the ensemble-mean surface ozone change is reproduced well. By using the Representative Concentration Pathway (RCP) emission scenarios it is shown how regional surface ozone is likely to respond to emission changes by 2050 and how changes in precursor emissions and atmospheric methane contribute to this. In the SRES A1B, A2 and B2 scenarios surface O_3 increases in 2050 there is little pollution control, whereas the RCP scenarios project stricter controls on precursor emissions. The study gave lower surface O_3 than at present. A large fraction of the difference between scenarios can be attributed to differences in methane abundance. The study showed the importance of limiting atmospheric methane growth, but also showed the uncertainty of modeled ozone responses to methane changes.

Global-scale NOx emission from aviation enhances ozone and reduces methane in the UTLS region. Skowron et al. [34] estimated that warming from ozone exceeded cooling from methane. This is in agreement with other studies of aircraft impact on RF from NOx initiated impact on ozone and methane ([24,33]). Methane reduction results in a small long-term reduction in tropospheric ozone (cooling) and a long-term reduction in water vapor in the stratosphere (cooling) from reduced oxidation of methane. Both have negative radiative forcing impacts.

Future impact of traffic emissions on atmospheric ozone and OH has been investigated separately for the three sectors: aircraft, maritime shipping and road traffic [32]. Results were presented from an ensemble of six different CTMs. The models simulated the atmospheric chemical composition in a possible high emission scenario (A1B), and with emissions from each of the transport sectors reduced by 5% to estimate sensitivities. The results were compared with more optimistic future emission scenarios (B1 and B1 ACARE). Current emissions are closer to the A1B than to the B1 scenario.

As a response to expected increase in emissions, air and ship traffic will increase their impacts on atmospheric O_3 and OH in the future, while the impact of road traffic is assumed to be reduced as a result of technological improvements. Summer maximum aircraft-induced O_3 occurs in the UTLS region at high latitudes, and could in 2050 be as high as 9 ppb for the zonal mean. Emissions from ship traffic have their largest O_3 impact in the maritime boundary layer with a maximum of 6 ppb over the North Atlantic Ocean during the summer months in 2050. The O_3 perturbations

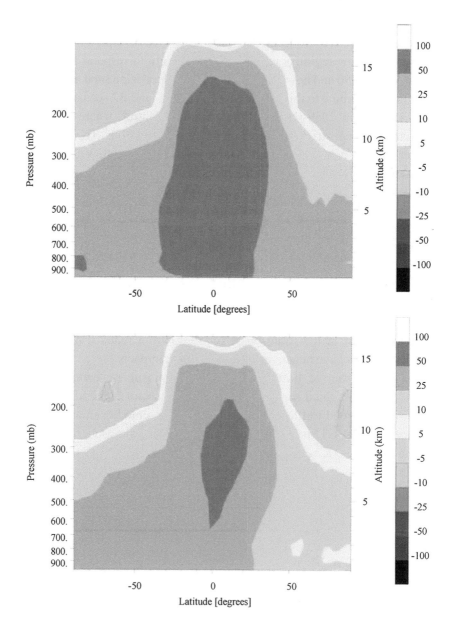

Figure 1. Percent change in the July zonally averaged ozone concentration between years 2000 and 2100, when the adopted change in the methane level (top) is included in the calculations and (bottom) is ignored. (Figure 1 is from Brasseur et al. [31])

of road traffic emissions are less pronounced than the perturbations from aircraft and ship. Maximum future impact of road is in the lower troposphere and peaks at 3 ppb over the Arabian Peninsula. However, for 2003, the emission assumed was much lower than the emission in 2000. A negative development in RF from road traffic prior to 2050 is temporary and induced by the strong decline in road emissions assumed. An emission scenario for road emissions (A1ACARE) assumes failures in the adopted B1 and A1B scenarios.

Calculations of NOx RF from ship have a negative overall impact from ozone and methane combined, while RF from aircraft and road NOx emissions are slightly positive. The RF from ship is estimated to become more negative in 2050 than in 2000.

Although the results vary between different model studies, the impact on RF is still positive if we consider all NOx effects from aircraft emissions. The Skowron et al. [34] paper, where six different aircraft NOx emission inventories were applied, gave rise to positive and negative RF impacts from air traffic. However, the variability of net radiative forcing impacts was significant between the inventories. From these calculations on aviation NOx, a wide range of Global Warming Potentials (GWP) for a 100-year time horizon was estimated. The estimates of vertical displacement of emission at chemically sensitive cruise altitudes strongly affects the assessment of the total radiative impact. In a study [40], it was also found that aviation NOx has an overall warming impact. Their study gave a GWP that was estimated to be 52 ± 52.

4.2.3 IMPACT OF INCREASED TEMPERATURES ON TROPOSPHERIC OZONE LEVELS

The increase in ozone production with increasing atmospheric temperatures was demonstrated for Western Europe for the summer months of 2003 in the paper by Solberg et al. [10] and for the Eastern Mediterranean region for the summer months of 2007 by Hodnebrog et al. [41]. The study by Solberg et al. [10] showed that an increase of atmospheric temperatures of 10 degrees, enhanced ozone levels by 4 ppb. They also argued that high temperatures triggered fires and isoprene emissions that will lead to more ozone production. Another positive link demonstrated in the papers was the strong relation of ozone levels to reduced uptake from enhanced surface dryness. Ozone levels above the ground were enhanced by 17 ppb when surface deposition was omitted.

Figure 2 shows that, during periods with high temperatures during the summer of 2007, ozone levels were higher than during the reference period 2000 to 2010. This is an indication that in a future warmer climate with more frequent heat waves, ozone levels are likely to be enhanced.

In order to show the impact of enhanced atmospheric temperatures on ozone levels we compare the observed diurnal maximum for each of the three summer months of 2007 with the reference period 2000–2010. We have also included in Table 1 the ozone values for the period 1961–1990. Two factors are important for increased ozone levels in recent years compared to earlier years. Enhanced emission of ozone precursors is a major factor for the increase in ozone levels as demonstrated in Table 1. We also suggest that increased frequency of heath waves due to climate change in the future will enhance pollution levels, including ozone, over regions with high emissions, as is shown for Western Europe in 2003 [10], and for the Eastern Mediterranean region in 2007 [41].

4.3 CLIMATE-CHEMISTRY INTERACTIONS

4.3.1 IPCC RELEVANCE

O3, its precursors and methane are affected by climate change through climate-chemistry interactions in the troposphere and in the stratosphere. Since ozone and methane are chemically active and important climate gases emission of ozone precursors and methane emission affect their contributions to climate change and their chemical behavior.

Modeling and observational analyses suggest a warmer climate degrades air quality (increasing surface ozone and particulate matter) in many populated regions. Such situations could be rather severe during pollution episodes. Although prior Intergovernmental Panel on Climate Change (IPCC) emission scenarios (SRES) had no restrictions on air

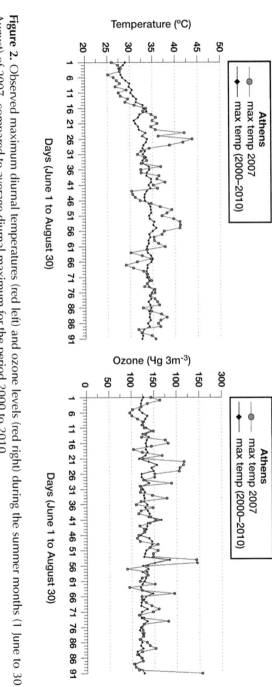

Figure 2. Observed maximum diurnal temperatures (red left) and ozone levels (red right) during the summer months (1 June to 30 August) of 2007, compared to average diurnal maximum for the period 2000 to 2010.

Table 1. Observed maximum monthly average of surface ozone (ppb) over Athens during June, July and August 2007, compared with observed monthly during the periods 2000–2010 and 1961–1990.

Maximum Monthly Average (Observations)	June	July	August	3 Months Average
2007	73	70	68	70
2000–2010	65	68	61	65
1961–1990	35	47	40	41

pollutants, current Representative Concentration Pathway (RCP) scenarios assume uniformly an extensive reduction in emissions of air pollutants. More recent estimates from the current generation of chemistry-climate models project improved air quality over the next century relative to those using the IPCC SRES scenarios. It is assumed that the two sets of projections likely bracket possible future emission scenarios. One finding is that uncertainties in emission-driven changes in air quality are generally greater than uncertainties in climate-driven changes.

Confidence in air quality projections is limited by the reliability of anthropogenic emission trajectories and the uncertainties in regional climate responses and feedback with the terrestrial biosphere, and oxidation pathways affecting ozone.

4.3.2 BIOSPHERE-ATMOSPHERE COUPLING

Changes in land cover may have significant consequences for atmospheric composition and air quality. Biogenic volatile organic compounds (VOCs; e.g., isoprene and monoterpenes) and nitric oxide (NO) emitted from certain vegetation species are important precursors for tropospheric ozone [3]. Although there have been several studies dealing with land use changes ([42,43,44,45]) possible effects on atmospheric chemistry and air pollution are still connected with significant uncertainties [46]. Wu et al. [46] studied the potential effects associated with future changes in vegetation driven by atmospheric CO_2 concentrations, climate, and anthropogenic land use over the 21st century. They performed a series of model experiments, which combined a general circulation model with a dynamic global vegetation model and an atmospheric chemical-transport model. Their studies indicate that climate-and CO_2-induced changes in vegetation composition and density between 2000 and 2100 could lead to decreases in summer afternoon surface ozone of up to 10 ppb over large areas of the northern mid-latitudes. This is largely driven by the substantial increases in ozone dry deposition associated with increases in vegetation density in a warmer climate with higher atmospheric CO_2 abundance. Climate-driven vegetation changes over the period 2000–2100 lead to general increases in isoprene emissions, globally by 15% in 2050 and 36% in 2100. These increases in isoprene emissions result in decreases in surface ozone concentrations where the NOx levels are low, such as in remote tropical rainforests. Over polluted regions, such as the north-eastern United States, ozone concentrations are calculated to increase with higher isoprene emissions in the future. For a future scenario with anthropogenic land use changes, Wu et al. [46] find less increase in global isoprene emissions due to replacement of higher-emitting forests by lower-emitting cropland. They find large regional variations in surface ozone toward 2100.

In a review article on ecosystems-atmosphere interactions and atmospheric composition change, Fowler et al. [3] included studies of ozone and methane. They were also considering a large number of gaseous and particle compounds (NO, NO_2, HONO, HNO_3, NH_3, SO_2, DMS, biogenic VOC and N_2O) that could affect ozone and methane processes and distributions. They found that changes in climate and chemical conditions could have a wide range of effects on the interaction of the biosphere with tropospheric ozone and hydroxyl radicals. Included in the study were ozone deposition, removal of NOx and biogenic emissions of ozone depleting compounds.

Changes in precursor emissions and long term changes in meteorology could affect tropospheric chemistry and the dry deposition process.

Characteristics of vegetation cover and land use may be altered on different scales as a result of human activities.

Higher future temperatures and changes in precipitation patterns could elevate CO_2 and O_3 concentrations, and may act as significant modifiers of surface uptake of gases. Sitch et al. [9] showed that uptake of CO_2 in vegetation is significantly less for high levels of ozone, although Kvalevåg and Myhre [18] got smaller reduction in the CO_2 uptake than in the previous study.

4.3.3 IMPACT OF METHANE EMISSIONS FROM PERMAFROST THAWING IN THE ARCTIC

Figure 3 shows an example of climate-chemistry interactions, which has the potential for non-linear effects on the atmospheric concentrations of methane, ozone and stratospheric water vapor, yielding further climate warming in a positive feedback loop. Methane is likely to be released from the Arctic as a result of strong future thawing of permafrost [47]. The release could be from conversion of organic carbon in the yedoma region where the organic content is high, 2% to 5% in the upper 25 m [48], or methane could be released from methane hydrates on the Arctic shelf during thawing in a warmer future climate ([29,49]). Zimov et al. [48] estimated that the yedoma region in Siberia contains 500 Gt of organic carbon and non-yedoma permafrost (excluding peatlands) contains another 400 Gt of organic carbon. Schuur et al. [50] estimate the carbon pool in permafrost areas to be 1672 Gt carbon (including peatlands). The deposited organic carbon is partly expected to be converted to methane (up to 30%) and released to the atmosphere after thawing [48]. There are some disagreements when it comes to the atmospheric impacts of methane deposited on the shallow continental shelf in permafrost regions, or conversion of organic carbon to methane. Shakhova et al. [49] assumes that methane could be rapidly released after a few years during thawing in the Arctic, while others argue that permafrost thawing will take much longer time ([19]).

Table 2. Modeled increase in the global release of CH_4, the tropospheric increase of ozone and stratospheric increase of water vapor, and the total radiative forcing for the selected atmospheric methane enhancements. Calculations are done with the Oslo chemistry transport model (Oslo CTM2).

Adopted Relative Increase in Atmospheric CH_4(%)	3.0	6.0	12.0
Increase in global release of CH_4 (Gt/year)	1.9	3.0	4.8
Maximum increase in tropospheric ozone (ppb)	50	70	120
Increase in stratospheric H_2O (%)	30–80	40–120	80–120
Increase in RF (W·m^{-2})	2.1	3.4	5.2

In Table 2 we estimate the additional amount of methane emissions, from permafrost thawing or other sources, that would be required to raise atmospheric methane to 3, 6 or 12 times the current concentrations. The required emissions are significantly higher than current methane emissions, which are approximately 0.5 Gt/year, but could be supplied by thawing Arctic soils. Such methane perturbations would also increase tropospheric ozone and stratospheric water vapor. The numbers for the ozone increases are for the gridbox with the highest ozone increase in the UTLS region (given in ppb), while water vapor increases are from the lower stratosphere (given in % increase). There are significant impacts of enhanced RF from permafrost release of methane. These numbers are based on the work by Isaksen et al. [29], and show that for large potential emissions of methane from permafrost thawing, we can expect significant climate impact, which represents a strong feedback in the climate system. Isaksen et al. [29] also found that large

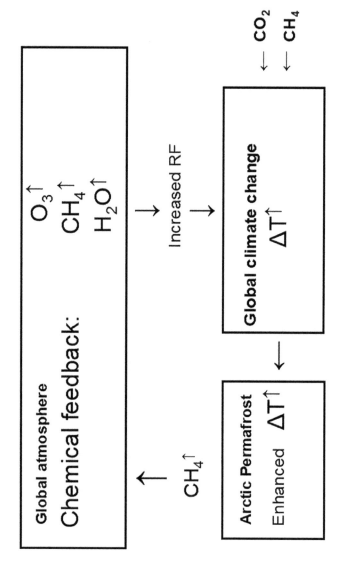

Figure 3. Schematic presentation of climate-chemistry interaction as a result of permafrost thawing and methane emissions.

permafrost emissions lead to increases in the methane lifetime through a significant impact of methane on its own lifetime, far beyond current feedback of 1.3 to 1.5.

4.3.4 IMPACT OF TRANSPORT ON THE LOW LATE-WINTER OZONE VALUES IN THE ARCTIC IN 2011

Arctic column ozone reached record low values (~230 DU) during March of 2011 ([11,12,13]) exposing the Arctic ecosystems to enhanced UV-B radiation. In the study by Isaksen et al. [13] ozone column north of 60 degrees N for the month of March in 2011 is given as 327 DU, compared with the range of average monthly values for the previous 10 years of 377 to 462 DU for the same region and the same month. The highest average monthly value for March in the Arctic was found in 2010. None of the previous 10 years had a similarly low column ozone during March in the Arctic as 2011. The study clearly showed that there are large year to year variations in late winter ozone columns over the Arctic. The cause of this anomaly was studied using the atmospheric Oslo CTM2 (chemistry transport model) driven by ECMWF meteorology. Simulations of Arctic ozone from 1997 to 2012 were performed, comparing parallel model runs with and without Arctic ozone chemistry between 1 January and 1 April in 2011. Even though there was considerable chemical loss of ozone in spring 2011 [51], a major part of the low ozone values observed over the Arctic in 2011 was dynamically driven [13]. Weakened transport of ozone from middle latitudes, at the same time as the polar vortex was strong, was the primary cause of the low ozone values. Pommereau et al. [52] have come to the same conclusion that reduced transport and a strong vortex were the main reason for the late winter low stratospheric ozone values in the Arctic in 2011. They also discussed the possibility for a relation to climate change, but did not find any indication of such connections.

Earlier studies on the effects of transport and chemistry on the inter-annual variability of ozone in the Arctic stratosphere for the period 1990 to 1998 [53] and for the period from 1992 to 2004 [54] concluded that both transport and chemistry contributed to ozone variability.

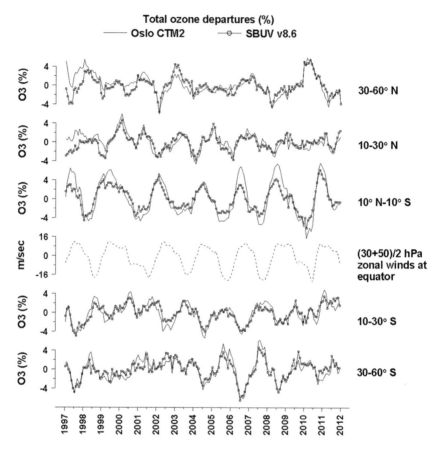

Figure 4. Column ozone variations, modeled (Oslo CTM2) and observed (SBUV) for different latitude zones between 1997 and 2012 (updated from [14]).

4.3.5 STRATOSPHERIC OZONE CHANGES DUE TO CHANGES IN DYNAMICS

We have studied how changes in dynamics affect ozone columns in the atmosphere [14]. Monthly mean ozone column from the chemical transport model Oslo CTM2 are compared with solar backscatter ultraviolet (SBUV) satellite observations (Figure 4). Ozone column values for different latitude zones in the Northern and Southern Hemispheres were compared. Ozone column variations from Oslo CTM2 are highly correlated with SBUV retrievals at all latitude zones. Equatorial zonal winds at 30 hPa were used as index to study the impact of quasi-biennial oscillation (QBO) on ozone. The impact of QBO was most pronounced at equatorial latitudes with amplitudes of +4% to −4%. At higher northern and southern latitudes, the amplitude is less pronounced and the oscillation phase lags that at the equator. We notice the lack of QBO effects at Southern Hemisphere mid-latitudes between about 2001 and 2004. We find that dynamics have a significant impact on the stratospheric ozone distribution as shown in this study. Seasonal variations in surface ozone and tropospheric ozone column calculated by the model are also presented in the study.

Tropospheric OH is clearly modified by stratospheric ozone columns and by QBO as shown in Figure 5, where modeled and observed ozone columns, observed UV-B radiation and modeled surface OH distribution for three Japanese stations are given for the period 1998 to 2012 [15] along with the equatorial zonal winds between 30 and 50 hPa. Column ozone variations, basically in the stratosphere, modify the shortwave solar radiation (UV-B radiation) penetrating to the troposphere. High ozone columns give reduced OH production in the troposphere, affecting methane lifetime and ozone production. We have also looked at high latitude stations in Canada. The relation between column ozone and tropospheric OH is not as clear at the stations in Canada.

4.4 CONCLUSIONS

We have demonstrated that there are possibilities for significant climate-chemistry interactions involving ozone and methane as key compounds. The interactions include the effect of temperature increases on chemical reactions, surface emissions of chemical compounds and dynamic changes affecting the ozone distribution in the stratosphere.

We have shown that during a period with enhanced summer temperatures over Athens in 2007 ozone levels were signficantly higher

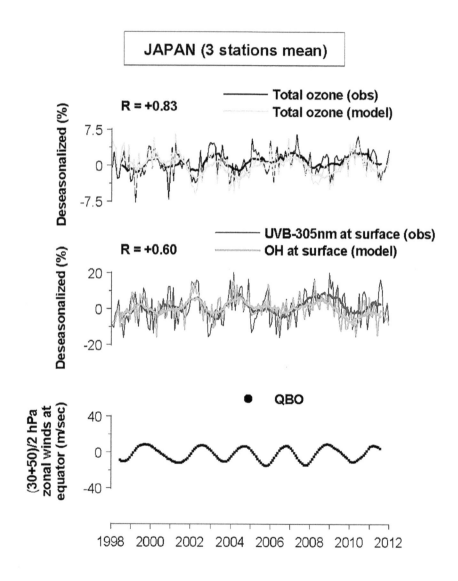

Figure 5. Column ozone variation (modeled and observed), tropospheric OH variation (modeled with Oslo CTM2) and observed tropospheric UV-B radiation for three Japanese stations, and observed zonal winds at 30 and 50 hPa at the Equator (updated from [15])

than the average for the same months during the period 2000 to 2010. The studies by Solberg et al. [10] and Hodnebrog et al. [41] demonstrated the link between increased atmospheric temperatures and enhanced ozone levels.

A future potential important area of climate-chemistry interactions is the release of methane from deposits under shallow ocean waters in the Arctic [49] or decomposition of organic deposits during Arctic thawing. Release of methane from the Arctic region could have impacts through enhancements of the atmospheric levels of methane itself, tropospheric ozone, stratospheric water vapor, and on radiative forcing, yielding a positive feedback in the climate system.

We have further demonstrated that we are able to represent well the observed column ozone change at different latitudinal bands, and that tropospheric OH variations are partly affected by ozone column variations [55]. The year-to-year variation in modeled ozone column distribution is large, basically showing the impact of transport, with some contribution from chemistry.

REFERENCES

1. IPCC. Climate Change 2007: The Physical Science Basis. Contribution of Working Group I to the Fourth Assessment Report of the Intergovernmental Panel on Climate Change; Solomon, S., Qin, D., Manning, M., Chen, Z., Marquis, M., Averyt, K.B., Tignor, M., Miller, H.L., Eds.; Cambridge University Press: Cambridge, UK/New York, NY, USA, 2007.
2. Isaksen, I.S.A.; Granier, C.; Myhre, G.; Berntsen, T.K.; Dalsøren, S.B.; Gauss, M.; Klimont, Z.; Benestad, R.; Bousquet, P.; Collins, W. Atmospheric composition change: Climate-Chemistry interactions. Atmos. Environ. 2009, 43, 5138–5192.
3. Fowler, D.; Pilegaard, K.; Sutton, M.A.; Ambus, P.; Raivonen, M.; Duyzer, J.; Simpson, D.; Fagerli, H.; Fuzzi, S.; Schjoerring, J.K. Atmospheric composition change: Ecosystems-Atmosphere interactions. Atmos. Environ. 2009, 43, 5193–5267.
4. Monks, P.S.; Granier, C.; Fuzzi, S.; Stohl, A.; Williams, M.; Akimoto, H.; Amman, M.; Baklanov, A.; Baltensperger, U.; Bey, I. Atmospheric composition change-global and regional air quality. Atmos. Environ. 2009, 43, 5268–5351.
5. Stevenson, D.S.; Johnson, C.E.; Collins, W.J.; Derwent, R.G.; Edwards, J.M. Future estimates of tropospheric ozone radiative forcing and methane turnover— The impact of climate change. Geophys. Res. Lett. 2000, 27, 2073–2076.

6. Fiore, A.M.; Naik, A.; Spracklen, D.V.; Steiner, A.; Unger, N.; Prather, M.; Bergmann, D.; Cameron-Smith, M.; Cionni, I.; Collins, W.J. Global air quality and climate. Chem. Soc. Rev. 2012, 41, 6663–6683.
7. Wild, O.; Fiore, A.M.; Shindell, D.T.; Doherty, R.M.; Collins, W.J.; Dentener, F.J.; Schultz, M.G.; Gong, S.; MacKenzie, I.A.; Zeng, G.; et al. Modelling future changes in surface ozone: A parameterized approach. Atmos. Chem. Phys. 2012, 12, 2037–2054.
8. National Academy of Sciences. Environmental Impact of Stratospheric Flight, Biological and Climatic Effects of Airceaft Emissions in the Stratosphere; National Academy of Sciences: Washington, DC, USA, 1975.
9. Sitch, S.; Cox, P.M.; Collins, W.J.; Huntingford, C. Indirect radiative forcing of climate change through ozone effects on the land-carbon sink. Nature 2007, 448, 791–794.
10. Solberg, S.; Hov, Ø; Søvde, A.; Isaksen, I.S.A.; Coddeville, P.; de Backer, H.; Forster, C.; Orsolini, Y.; Uhse, K. European surface ozone in the extreme summer 2003. J. Geophys. Res.: Atmos. 2008.
11. Manney, G.L.; Santee, M.L.; Rex, M.; Livesey, N.J.; Pitts, M.C.; Veefkind, P.; Nash, E.R.; Wohltmann, I.; Lehmann, R.; Froidevaux, L.; et al. Unprecedented Arctic ozone loss in 2011. Nature 2011, 478, 469–475.
12. Sinnhuber, B.-M.; Stiller, G.; Ruhnke, R.; von Clarmann, T.; Kellmann, S.; Aschmann, J. Arctic winter 2010/2011 at the brink of an ozone hole. Geophys. Res. Lett. 2011.
13. Isaksen, I.S.A.; Zerefos, C.; Wang, W.-C.; Balis, D.; Eleftheratos, K.; Rognerud, B.; Stordal, F.; Berntsen, T.K.; LaCasce, J.H.; Søvde, O.A.; et al. Attribution of Arctic ozone loss in March 2011. Geophys. Res. Lett. 2012.
14. Eleftheratos, K.; Isaksen, I.S.A.; Zerefos, C.; Nastos, P.; Tourpali, K.; Rognerud, B. Ozone variations derived by a chemical transport model. Water Air Soil Pollut. 2013.
15. Zerefos, C.S.; Tourpali, K.; Eleftheratos, K.; Kazadzis, S.; Meleti, C.; Feister, U.; Koskela, T.; Heikkilä, A. Evidence of a possible turning point in solar UV-B over Canada, Europe and Japan. Atmos. Chem. Phys. 2012, 12, 2469–2477.
16. Hauglustaine, D.A.; Lathière, J.; Szopa, S.; Folberth, G.A. Future tropospheric ozone simulated with a climate-chemistry-biosphere model. Geophys. Res. Lett. 2005.
17. Liao, H.; Chen, W.-T.; Seinfeld, J.H. Role of climate change in global predictions of future tropospheric ozone and aerosols. J. Geophys. Res.: Atmos. 2006.
18. Kvalevåg, M.M.; Myhre, G. The effect of carbon-nitrogen coupling on the reduced land carbon sink caused by tropospheric ozone. Geophys. Res. Lett. 2013, 40, 3227–3231.
19. Nisbet, E.G.; Chappellaz, J. Shifting Gear, Quickly. Science 2009, 324, 477–478.
20. Isaksen, I.S.A.; Gauss, M.; Myhre, G.; Walter Anthony, K.M.; Ruppel, C. Strong atmospheric chemistry feedback to climate warming from Arctic methane emissions. Glob. Biogeochem. Cycles 2011.
21. Schmidt, G.A.; Shindell, D.T. Atmospheric composition, radiative forcing, and climate change as a consequence of massive methane relies from gas hydrates. Paleoceanography 2003.
22. Shakhova, N.; Semiletov, I.; Leifer, I.; Sergienko, V.; Salyuk, A.; Kosmach, D.; Chernykh, D.; Stubbs, C.; Nicolsky, D.; Tumskov, V.; Gustafsson, Ö. Ebullition and

storm-induced methane release from the East Siberian Arctic shelf. Nat. Geosci. 2014, 7, 64–70.

23. Eyring, V.; Isaksen, I.S.A.; Berntsen, T.; Collins, W.J.; Corbett, J.J.; Endresen, O.; Grainger, R.G.; Moldanova, J.; Schlager, H.; Stevenson, S. Transport impacts on atmosphere and climate: Shipping. Atmos. Environ. 2010, 44, 4735–4771.

24. Hoor, P.J.; Borken-Kleefeld, D.; Caro, O.; Dessens, O.; Endresen, M.; Gauss, V.; Grewe, D.; Hauglustaine, I.S.A.; Isaksen, P.; Jöckel, J.; et al. The impact of traffic emissions on atmospheric ozone and OH: Results from QUANTIFY. Atmos. Chem. Phys. 2009, 9, 3113–3136.

25. IPCC. Atmospheric Chemistry and Greenhouse Gases; IPCC WGI Third Assessment Report. IPCC: Geneva, Switzerland, 2001; pp. 239–287.

26. Isaksen, I.S.A.; Hov, Ø. Calculation of trends in the tropospheric concentration of O3, OH, CO, CH4 and NOx. Tellus 1987, 39B, 271–285.

27. Prather, M.; Ehhalt, D. Atmospheric chemistry and greenhouse gases. In Climate Change 2001: The Scientific Basis; Houghton, J.T., Ding, Y., Griggs, D.J., Noguer, N., van der Linden, P.J., Xiaosu, D., Maskell, K., Johnson, C.A., Eds.; Cambridge University Press: Cambridge, UK, 2001; pp. 239–287.

28. Voulgarakis, A.; Naik, V.; Lamarque, J.-F.; Shindell, D.T.; Young, P.J.; Prather, M.J.; Wild, O.; Field, R.D.; Bergmann, D.; Cameron-Smith, P.; et al. Analysis of present day and future OH and methane lifetime in the ACCMIP simulations. Atmos. Chem. Phys. 2013, 13, 2563–2587.

29. Mickley, L.J.; Jacob, D.J.; Rind, D. Uncertainty in preindustrial abundance of tropospheric ozone: Implications for radiative forcing calculations. J. Geophys. Res.: Atmos. 2001, 106, 3389–3399.

30. Fiore, A.M.; Levy, H., II; Ming, Y.; Fang, Y.; Horowitz, L.W. Interactions between air quality and climate. In Air Pollution Modeling and Its Application XX; Springer: Dordrecht, The Netherlands, 2010; pp. 481–489.

31. Brasseur, G.P.; Schultz, M.; Granier, C.; Saunois, M.; Diehl, T.; Botzet, M.; Roeckner, E.; Walters, S. Impact of climate change on the future chemical composition of the global troposphere. J. Clim. 2006, 19, 3932–3951.

32. Hodnebrog, O.; Berntsen, T.K.; Dessens, O.; Gauss, M.; Grewe, V.; Isaksen, S.A.; Koffi, B.; Myhre, G.; Olivie, D.; Prather, M.J.; et al. Future impact of traffic emissions on atmospheric ozone and OH based on two scenarios. Atmos. Chem. Phys. 2012, 12, 12211–12225.

33. Hoyle, C.R.; Myhre, G.; Isaksen, I.S.A. Present-day contribution of anthropogenic emissions from China to the global burden and radiative forcing of aerosol and ozone. Tellus 2009, 61B, 618–626.

34. Skowron, A.; Lee, D.S.; de Leon, R.R. The assessment of the impact of aviation NOx on ozone and other radiative forcing responses—The importance of representing cruise altitudes accurately. Atmos. Environ. 2013, 74, 159–168.

35. Søvde, O.A.; Gauss, M.; Smyshlyaev, S.P.; Isaksen, I.S.A. Evaluation of the chemical transport model Oslo CTM2 with focus on arctic winter ozone depletion. J. Geophys. Res.: Atmos. 2008.

36. Søvde, O.A.; Prather, M.J.; Isaksen, I.S.A.; Berntsen, T.K.; Stordal, F.; Zhu, X.; Holmes, C.D.; Hsu, J. The chemical transport model Oslo CTM3. Geosci. Model Dev. 2012, 5, 1441–1469.

37. Tsai, I.-C.; Chen, J.-P.; Lin, P.-Y.; Wang, W.-C.; Isaksen, I.S.A. Sulfur cycle and sulfate radiative forcing simulated from a coupled global climate-chemistry model. Atmos. Chem. Phys. 2010, 10, 3693–3709.

38. Unger, N.; Shindell, D.T.; Koch, D.M.; Amann, M.; Cofala, J.; Streets, D.G. Influences of man-made emissions and climate changes on tropospheric ozone, methane, and sulfate at 2030 from a broad range of possible futures. J. Geophys. Res.: Atmos. 2006.

39. Ødemark, K.; Dalsøren, S.B.; Samset, B.H.; Berntsen, T.K.; Fuglestvedt, J.S.; Myhre, G. Short-lived climate forcers from current shipping and petroleum activities in the Arctic. Atmos. Chem. Phys. 2012, 12, 1979–1993.

40. Holmes, C.D.; Tang, Q.; Prather, M.J. Uncertainties in climate assessment for the case of aviation NO. Proc. Natl. Acad. Sci. USA 2011, 108, 10997–11002.

41. Hodnebrog, Ø.; Solberg, S.; Stordal, F.; Svendby, T.M.; Simpson, D.; Gauss, M.; Hilboll, A.; Pfister, G.G.; Turquety, S.; Richter, A.; et al. Impact of forest fires, biogenic emissions and high temperatures on the elevated eastern mediterranean ozone levels during the hot summer of 2007. Atmos. Chem. Phys. 2012, 12, 8727–8750.

42. Ganzeveld, L.; Bouwman, L.; Stehfest, E.; van Vuuren, D.P.; Eickhout, B.; Lelieveld, J. Impact of future land use and land cover changes on atmospheric chemistry-climate interactions. J. Geophys. Res.: Atmos. 2010.

43. Sanderson, M.G.; Jones, C.D.; Collins, W.J.; Johnson, C.E.; Derwent, R.G. Effect of climate change on Isoprene emissions and surface ozone levels. Geophys. Res. Lett. 2003.

44. Turner, B.L., II; Meyer, W.B.; Skole, D.L. Global landuse/land cover change: Towards an integrated study. Ambio 1994, 23, 91–95.

45. Turner, B.L., II; Skole, D.; Sanderson, S.; Fischer, G.; Fresco, L.; Leemans, R. Land-Use and Land-Cover Change Science/Research Plan; IHDP Report No. 7. Royal Swedish Academy of Sciences: Stockholm, Sweden, 1995.

46. Wu, S.; Mickley, L.J.; Kaplan, J.O.; Jacob, D.J. Impacts of changes in land use and land cover on atmospheric chemistry and air quality over the 21st century. Atmos. Chem. Phys. 2012, 12, 1597–1609.

47. Shakhova, N.E.; Sergienko, V.I.; Semiletov, I.P. The contribution of the East Siberian shelf to the modern methane cycle. Her. Russ. Acad. Sci. 2009, 79, 217–246.

48. Zimov, S.A.; Schuur, E.A.G.; Chapin, F.S., III. Permafrost and global carbon budget. Science 2006, 312, 1612–1613.

49. Shakhova, N.E.; Alekseev, V.A.; Semiletov, I.P. Predicted methane emission on the east Siberian shelf. Dokl. Earth Sci. 2010, 430, 190–193.

50. Schuur, E.A.G.; Bockheim, J.; Candell, J.G.; Auskirchen, E.; Field, C.B.; Goeyachkin, S.V.; Hagemann, S.; Kuhry, P.; Lafleur, P.M.; Lee, H.; et al. Vunerability of permafrost carbon to climate change: Implications for the carbon cycle. Bioscience 2008, 58, 701–714.

51. Kuttippurath, J.; Godin-Beekmann, S.; Lefèvre, F.; Nikulin, G.; Santee, M.L.; Froidevaux, L. Record-breaking ozone loss in the Arctic winter 2010/2011: Comparison with 1996/1997. Atmos. Chem. Phys. 2012, 12, 7073–7085.

52. Pommereau, J.-P.; Goutail, F.; Lefevre, F.; Pazmino, A.; Adams, C.; Dorokhov, V.; Eriksen, P.; Kivi, R.; Stebel, K.; Xhou, X.; et al. Why unpresedented ozone loss in

the Arctic in 2011? Is it related to climate change? Atmos. Chem. Phys. 2013, 13, 5299–5308.

53. Chipperfield, M.P.; Jones, R.L. Relative influences of atmospheric chemistry and transport on Arctic ozone trends. Nature 1999, 400, 551–554.

54. Tegtmeier, S.; Rex, M.; Wohltmann, I.; Krüger, K. Relative importance of dynamical and chemical contributions to Arctic wintertime ozone. Geophys. Res. Lett. 2008.

55. Eleftheratos, K.; Zerefos, C.S.; Gerasopoulos, E.; Isaksen, I.S.A.; Rognerud, B.; Dalsoren, S.; Varotsos, C. A note on the comparison between total ozone from Oslo CTM2 and SBUV satellite data. Int. J. Remote Sens. 2011, 32, 2535–2545.

CHAPTER 5

The Role of HFCs in Mitigating 21st Century Climate Change

Y. XU, D. ZAELKE, G. J. M. VELDERS, AND V. RAMANATHAN

5.1 INTRODUCTION

The ozone depleting substances (ODSs) (e.g., chlorofluorocarbons (CFCs), hydrochlorofluorocarbons (HCFCs)), halons, and HFCs are part of a family of gases known as halocarbons. Halocarbons are used as refrigerants, propellants, cleaning and foam blowing agents, and fire extinguishers, etc. Molina and Rowland (1974) identified the potent stratospheric ozone depleting effects of CFCs. This was followed, within a year, by the discovery of the potent greenhouse effect of the halocarbons CFC-11 and CFC-12 (Ramanathan, 1975). Many studies confirmed this finding and estimated that the global warming potential (GWP) of CFC-11 and CFC-12 (using a 100 yr time horizon) at 4750 and 10 900 respectively, as summarized by the Intergovernmental Panel on Climate Change Fourth

Assessment Report (Forster et al., 2007). Ramanathan (1975) set the stage for identifying numerous other non-CO_2 greenhouse gases (GHGs) in the atmosphere such as CH_4 and O_3 among others (seeWang et al., 1976 and Ramanathan et al., 1985a). The first international assessment of the climate effects of non-CO_2 gases was conducted in 1985 (Ramanathan et al., 1985b) and it concluded that CO_2 was the dominant contributor to greenhouse forcing until 1950s, and since the 1960s non-CO_2 gases have begun to contribute as much as CO_2. A more recent list of the non-CO_2 GHGs can be found in Pinnock et al. (1995) and Forster et al. (2007).

Most of HFCs now in use, along with CH_4, O_3, and BC (black carbon aerosols), have relatively short lifetimes in the atmosphere in comparison with long-lived GHGs, such as CO_2 and N_2O (nitrous oxide) (e.g., see Smith et al., 2012), and are therefore referred to as short-lived climate pollutants (SLCPs). The lifetime of BC is several days to weeks, tropospheric O_3 is a few months, and CH_4 is about 12 yr. The global average lifetime, weighted by the production of the various HFCs now in commercial use, is about 15 yr, with a range of 1 to 50 yr (Table 1). Because the lifetimes of the SLCPs are much shorter than that of CO_2, a significant portion of which remains in the atmosphere for centuries to millennia, the radiative forcing by SLCPs will decrease significantly within weeks to a few decades after their emissions are reduced.

Motivated by modeling studies (e.g., Ramanathan and Xu, 2010; Shindell et al., 2012), policy makers are showing increasing interest in fast-action climate mitigation strategies that target SLCPs (Wallack and Ramanathan, 2009; Molina et al., 2009). Ramanathan and Xu (2010) (hereafter RX10) concluded that as much as 0.6°C warming can be avoided by mid-21st century using current technologies to reduce all four SLCPs, with mitigation of HFCs contributing about 20% (0.1 C) to the avoided warming by 2050. Furthermore, RX10 also showed that exceeding the 2 ° C warming threshold can be delayed by three to five decades beyond 2050 by these efforts. Based on an international assessment commissioned by the United Nations Environment Programme (UNEP) and the World Meteorological Organization (WMO) (UNEP and WMO, 2011), Shindell et al. (2012) used a 3-dimensional climate model to account for reductions in CH_4, O_3, and BC emissions (but not HFCs) using mitigation scenarios

Table 1. Indicative applications using HFCs*.

HFC	Indicative applications UNEP and WMO (2011)	Lifetime WMO (2011)	100 yr GWPWMO (2011)
HFC-134a	Mobile and stationary air conditioning and refrigeration, foams, medical aerosols and cosmetic and convenience aerosol products	13.4	1370
HFC-32	In blends for refrigeration and air conditioning	5.2	716
HFC-125	In blends for refrigeration and air conditioning	28.2	3420
HFC-143a	In blends for refrigeration and air conditioning	47.1	4180
HFC-152a	Foams, aerosol products	1.5	133
HFC-227ea	Foams, medical aerosols, fire protection	38.9	3580
HFC-245fa	Foams	7.7	1050
HFC-365mfc	Foams	8.7	842
HFC-43-10mee	Solvents	16.1	1660

* HFC-23 is not included in the scenarios discussed here. Although it is currently the second most abundant HFC in the atmosphere, it is assumed that the majority of this chemical is produced as a byproduct of HCFC-22 production and not because of its use as a replacement for CFCs and HCFCs. Hence, the emissions of HFC-23 depend on a different set of assumptions than the other HFCs (Velders et al. 2009). Miller and Kuijpers (2011) estimated that HFC-23 emissions increase could contribute $0.014 Wm^{-2}$ to radiative forcing in 2050. Therefore, the contributed warming due to potential HFC-23 will be only about 0.01 °C by our estimation.

similar to those employed in RX10. UNEP and WMO (2011) as well as Shindell et al. (2012) calculated the avoided warming to be 0.5(±0.05) °C by 2070. This estimate is consistent with RX10, which would also yield 0.5°C avoided warming if only CH_4, O_3, and BC were mitigated. All three studies calculated that full implementation of mitigation measures for these three SLCPs can reduce the rate of global warming during the next several decades by nearly 50 %. Furthermore, Arctic warming can be reduced by two-thirds over the next 30 yr compared to business as usual (BAU) scenarios (UNEP and WMO, 2011).

However, with the exception of the RX10 study, HFCs have thus far not been included in analyses of the temperature mitigation benefit from SLCP mitigation. Even RX10 did not recognize the full potential of the radiative forcing increase, as shown recently by Velders et al. (2012), due to an unconstrained use of HFCs toward the end of this century. Therefore, what has been missing in the previous studies is the potentially large increase in HFC use. The present study builds upon RX10 to further account for the newly developed projections of HFC emissions and provides a detailed analysis of the implication of HFC mitigation on global temperature.

5.2 METHODS

5.2.1 HFC EMISSION PROJECTION

Because of their catalytic destruction of stratospheric ozone, production and consumption of CFCs, HCFCs and other ODSs are being phased out under the Montreal Protocol (Andersen and Sarma, 2002; Andersen et al., 2007). With the phase-out of CFCs under the Montreal Protocol completed in 1996 in developed countries and in 2010 in developing countries (UNEP, 2010), and with the scheduled phase-out of HCFCs by 2030 in developed countries, and 2040 in developing countries (UNEP, 2007), HFCs are increasingly being used as alternatives in applications that traditionally used CFCs, HCFCs and other ODSs to meet much of the demand for refrigeration, air conditioning, heating and thermalinsulating foam production (Velders et al., 2012). HFCs do not destroy the ozone layer (Ravishankara et al., 1994) but are potent GHGs (Velders et al., 2009).

The demand for HFCs is expected to increase in both developed and developing countries, especially in Asia, in the absence of regulations, as is the demand for HCFCs for feedstock (Velders et al., 2009). HFCs are the fastest growing GHGs in the US, where emissions grew nearly 9% between 2009 and 2010 compared to 3.6% for CO_2 (EPA, 2012). Globally,

HFC emissions are growing 10 to 15% per year and are expected to double by 2020 (WMO, 2011; Velders et al., 2012). The presence of HFCs in the atmosphere results almost completely from their use as substitutes for ODSs (Table 1).

The future HFC projection in this study is estimated using (1) the growth rates of gross domestic product (GDP) and populations from the Special Report on Emissions Scenarios (SRES) (IPCC, 2000), and (2) the replacement patterns of ODSs with HFCs and not-in-kind technologies as observed in the past years in Western countries. We assumed that these replacement patterns will stay constant and will hold for developing countries. The European Union regulation (842/2006) aimed at moving away from high-GWP HFCs is also included in the HFC projections used here. Readers are referred to Velders et al. (2009) for more details in HFC scenario development (e.g., emissions by substance, region and years). Because the projected forcing from HFC-23 is much smaller than that from intentionally produced HFCs, it is not included in this study. In spite of potential large increases in HFC-23 from the continued production of HCFC-22 for feedstock, the HFC-23 forcing in 2050 is $0.014 \, \text{Wm}^{-2}$ (Miller and Kuijpers, 2011) and the associated warming is only about 0.01°C.

5.2.2 OTHER EMISSION PROJECTION

The future emission scenarios of CO_2 are adopted from the Representative Concentration Pathway (RCP, van Vuuren et al., 2011) database. We take RCP 2.6 (van Vuuren et al., 2007) as mitigation case and RCP 6.0 (Hijioka et al., 2008) as BAU case for CO_2. CO_2 emissions in the mitigation case will decline by half in mid-21st century, while the BAU CO_2 emissions are projected to continue to increase until 2080. The peak CO_2 atmospheric concentration is 660 and 440 ppm under BAU and mitigation cases, respectively. We note that CO_2 scenarios under RCP 6.5 and 2.6 may have different assumptions with regard to emission sectors and therefore the difference between those two pathways may not directly represent the effect of mitigation efforts. The SLCP projections, except for HFCs, are retained from RX10. Under a BAU scenario, CH4 emissions are predicted to rise by 40% in 2030, and BC emissions are projected to increase by 15% by 2015 and then level off. The mitigation scenarios follow recommendations from studies by the International Institute for Applied Systems Analysis (IIASA) (Cofala et al., 2007) and the Royal Society (2008) that maximum feasible reductions of air pollution regulations can result in reductions of 50% in CO emissions and 30% in CH_4 emissions by 2030, as well as reductions of 50% in BC emissions by 2050.

5.2.3 MODELS

The model used in RX10 is an integrated carbon and radiant energy balance model. It adopts the Bern CO_2 geochemistry model (Joos et al., 1996) to estimate the atmospheric CO_2 concentration from emissions. The model links emissions of pollutants with their atmospheric concentrations and the change in the radiative forcing. The carbon-geochemistry model is then integrated with an energy balance climate model with a 300m ocean mixed layer and a climate sensitivity of 0.8 (0.5 to 1.2)°C/(Wm^{-2}) to simulate the temporal evolution of global mean surface temperature. The model also accounts for historical variations in the global mean radiative forcing to the system attributable to natural factors, GHGs and air pollutants including sulfates, nitrates, carbon monoxide, ozone, BC, and organic carbons. The model is capable of simulating the observed historical temperature variations (Fig. 2), as well as the historical CO_2 concentration and ocean heat content (Box 1 in RX10).

5.3 RESULTS AND DISCUSSIONS

5.3.1 LARGE INCREASE OF HFC FORCING

The radiative forcing of HFCs in 2008 was small at less than 1% of the total forcing from long-lived GHGs (WMO, 2011). However, without fast action to limit their growth, the radiative forcing of HFCs could increase from nearly 0.012 Wm^{-2} in 2010 to up to 0.4 Wm^{-2} in 2050 (BAU high in

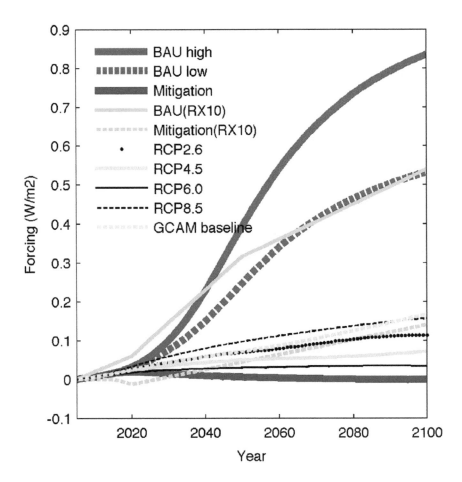

Figure 1. HFC radiative forcing change (Wm^{-2}) since the year of 2005. Note we include both upper (solid line) and lower limits (dash line) of HFC growth under BAU scenarios. The scenarios previously adopted in RX10 and from various other sources are shown for reference.

Fig. 1). The 0.4 Wm^{-2} is equal to nearly 30 to 45% of CO_2 forcing increase by 2050 (if CO_2 follows BAU and mitigation scenarios, respectively; see Sect. 2.2 for scenario descriptions), or about the same forcing contributed by current CO_2 emissions from the transportation sector (IEA, 2011). In the scenarios discussed here, the demand for HFCs for the period 2050 to 2100 is assumed to maintain at the 2050 levels (assuming complete market saturation), which results in increasing HFC abundances and radiative forcing past 2050, with HFC forcing possibly reaching as high as $0.8 Wm^{-2}$ in 2100 (BAU high in Fig. 1).

We also calculate HFC forcing from emission data provided by the RCP database and compare them with our forcing projections. Similar comparisons of future emission and forcing are also shown in Fig. 5.5 of theWMO(2011) assessment. However, a direct comparison is difficult, because with the exception of RCP 8.0 (Riahi et al., 2007), the scenarios include various mitigation policies as assumptions and therefore cannot be considered as "BAU" scenarios. The Global Change Assessment Model (GCAM) group that produces RCP 4.5 (Wise et al., 2009; Thomson et al., 2011) does, however, make available a "BAU" scenario (GCAM baseline in Fig. 1), which does not include explicit mitigation actions (Smith et al., 2011). As a result, HFC forcing is two times larger in the GCAM baseline scenario than the associated RCP 4.5 scenario. The HFC BAU projections used in this study are substantially higher over the long-term than the projections of other studies, including RCP 8.0 and GCAM baseline. In the GCAM baseline, for example, HFC forcing increase from 2005 is less than $0.2 Wm^{-2}$ in 2100, as compared to 0.5 to $0.8 Wm^{-2}$ in our BAU scenarios.

There are several reasons for the discrepancies in HFC projections. (1) HFC scenarios have not received much attention in the development of the RCP database. Individual integrated assessment modeling groups have adopted various assumptions and techniques in developing those RCP projections and associated reference scenarios. However, detailed descriptions of HFC projection in RCPs and associated reference scenarios are not available in relevant papers, which are more focused on long-lived GHGs. (2) Most, if not all, of the RCP scenarios for HFCs were developed before 2007. Therefore, they did not take into account the accelerated HCFC phase-out in both developing and developed countries agreed by the parties at the 19th Meeting of the Parties to the Montreal Protocol in September 2007, which will lead to lower future HCFC emissions and higher HFC emissions (Meinshausen et al., 2011). (3) At least some RCP

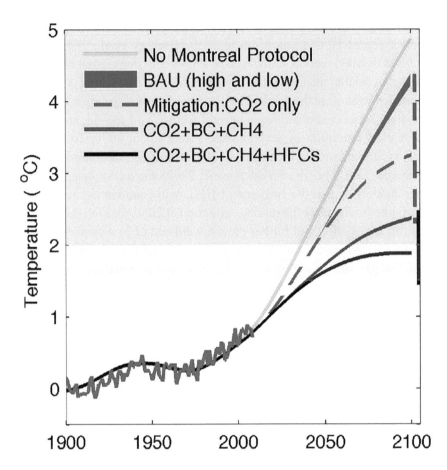

Figure 2. Model simulated temperature change under various mitigation scenarios that include CO_2 and SLCPs (BC, CH_4, HFCs). BAU case (red solid line with spread) considers both high and low estimates of future HFC growths (as shown in red solid and dash lines in Fig. 1). Note this uncertainty of temperature projection related to HFC scenarios is around 0.15 C at 2100. The vertical bars next to the curve show the uncertainty of temperature projection at 2100 due to climate sensitivity uncertainty. For simplicity, we only show the cases of CO_2 mitigation (red dash line) and full mitigation (black line).

scenarios did not take into account the large observed growth in HFC use and atmospheric concentrations since 2000 (WMO, 2011). The linear growth in RCP scenarios (Fig. 1 and also see Fig. 3.22 of Clarke et al.,

2007) is distinctly different from the assumptions of a growing market in developing countries, which was the basis of our HFC scenario. (4) Finally, another recent HFC scenario up to 2050 (Gschrey et al., 2011), which includes detailed market analysis, also shows emissions much higher than RCPs, but smaller than ours. The smaller emissions in Gschrey et al. (2011) compared to ours are the result of two assumptions: first, a larger fraction of non-fluorocarbon alternatives in several sectors; and second, and second, the market saturation in several sectors after about 2030, 10 yr earlier than in our scenarios, which considered consumption saturation for a few sectors at 2040 (Fig. 1b of Velders et al., 2009) and a complete saturation after 2050. Note that the emission of HFCs will continue increasing for a short time period after the market saturation at 2050, and the mixing ratio and forcing of HFC will further grow toward end of 21st century (Fig. 1). We acknowledge that some RCP models project HFC in a detailed method (e.g., by gas and sector based on the evolution of multiple drivers over time including vehicle demand, building airconditioning use, etc.). Some RCP models and Gschrey et al. (2011) have also assumed some emission drivers do not scale with GDP over the long-term due to saturation effects (e.g., floor space of buildings), which could be even more important on a century timescale. The differences in considering saturation effects may be a reason that our projections yield larger emissions.

In conclusion, differences in HFC scenarios arise from large differences in their underlying assumptions and to the level they take into account recent information. HFC projections in some RCP scenarios do not use the more recent information as in Velders et al. (2009) and Gschrey et al. (2011). The scenarios in Velders et al. (2009) are based on assumptions similar to those of IPCC-SRES with respect to growth rates in GDP and population, but have incorporated new current information on (1) reported recent increases in consumption of HCFCs in Article 5 (developing) countries of about 20% per year (through 2007), (2) replacement patterns of HCFCs by HFCs as reported in non-Article 5 (developed) countries, and (3) accelerated phase-out schedules of HCFCs in Article 5 and non-Article 5 countries (2007 Adjustment of the Montreal Protocol).We note that this HFC scenario is not necessarily a more accurate forecast of future HFC emissions than other scenarios, but a projection of what can happen if developed countries continue current practices in replacing ODSs with HFCs and if developing countries follow this path as well.

In contrast to the large increase under BAU scenarios, replacing those HFCs currently in use with low-GWP HFC alternatives that have lifetimes of less than one month can eliminate future HFC forcing increase (Velders et al., 2012). Under the mitigation scenario, the total HFC radiative forcing in 2050 would be less than its current value (Mitigation in Fig. 1). Alternatives with no direct impact on climate, including ammonia, carbon dioxide, and hydrocarbons, as well as low-GWP HFCs and not-in-kind alternatives, are already in commercial use in a number of sectors. For other sectors, alternatives are being evaluated or further developed (UNEP, 2011). The calculation of climate mitigation assumes that the selected alternatives do not compromise energy efficiency, an assumption that appears reasonable in light of the historic trend of increased energy efficiency when chemicals are phased out under the Montreal Protocol (Andersen and Morehouse, 1997; Andersen and Sarma, 2002; Andersen et al., 2007).

5.3.2 IMPLICATION FOR GLOBAL TEMPERATURE

The simulated temperature trends (Fig. 2) agree with the earlier studies (Shindell et al., 2012; UNEP and WMO, 2011) that combined mitigation of CH_4, BC and O_3 can mitigate 0.5°C of warming by the mid-century. It also agrees with RX10 that HFCs contribute about 0.1°C to the avoided warming of 0.6°C by 2050 and that SLCPs are critical for limiting the warming below 2°C. CO_2 mitigation, although begun in 2015, has very little effect for the near-term (see difference between red solid line and red dash line in Fig. 2). Focusing on the longer timescale of the end of the century (Fig. 2), CO_2 mitigation plays a major role in reducing additional warming by as much as 1.1°C by 2100. Next, the combined measures (CO_2, CH_4, BC and O_3) considered in UNEP and WMO (2011) and Shindell et al. (2012) are not sufficient to limit the warming below 2°C (blue line in Fig. 2), had these studies included the updated projected HFC growth patterns of Velders et al.

(2009) in their BAU scenarios. Mitigation of the potential growth of HFCs is shown to play a significant role in limiting the warming to below 2°C and could contribute additional avoided warming of as much as 0.5°C by 2100 (blue and black line in Fig. 2). Using the lower limits of BAU increase of HFC (red dash line in Fig. 1), 0.35°C warming will be avoided.

The results are consistent with RX10 for the near-term, but the avoided warming from HFCs towards the end of the century is 100% higher in this study, due to the updated forcing scenarios accounting for the high HFC growth rate (green lines in Fig. 1 for a comparison with RX10 forcing scenarios). Replacing HFCs with available low-GWP substitutes that have a lifetime of one month or less, or with other materials or technologies, can provide up to 0.35 to 0.5°C of warming mitigation by 2100 in the scenarios used here. The important point to note is that assessing the role of HFCs in climate change depends on what BAU (i.e., reference/ baseline) scenarios the climate models assume for HFCs in their simulations. Many climate models assume much smaller growth of HFC emission, because of the implicit assumption that replacements with low impact on climate for high-GWP HFCs will be adopted extensively during this century, an assumption that largely depends on the extent of policy interventions, as well as technological and economic developments. Our study, however, shows that if current growth rates of high-GWP HFCs continue, the additional warming from HFCs alone will be as much as 0.5°C during this century. The potential temperature mitigation by the end of this century, from HFC replacement, is in addition to the 1°C potential mitigation from other SLCP reductions (Fig. 2; also see RX10). When mitigation effort to reduce high-GWP HFCs is combined with that on BC and CH_4, 0.6°C warming can be avoided by 2050 and 1.5°C by 2100 (black solid line vs. red dash line in Fig. 2). This would cut the cumulative warming since 2005 by 50% at 2050 and by 60% at 2100 from the corresponding CO_2-only mitigation scenarios (red dash line in Fig. 2). Based on our high HFC growth scenarios, the contribution to the avoided warming at 2100 due to HFC emission control is about 40% of that due to CO_2 emission control. Considering the nearterm (2050) timescale, HFC emission is even more effective (140% of CO_2 mitigation) in curbing the warming. Given the limited knowledge regarding climate sensitivity (0.5 to 1.2°C/(Wm^{-2})), the absolute value of projected temperature at the end of 21st century is also uncertain (vertical bars in Fig. 2), but the relative contribution of HFC to reducing the warming is still significant and less subject to such uncertainty.

5.4 CONCLUSIONS

The concept of "short-lived climate pollutants" highlights the shorter lifetime of those pollutants (including HFCs) as compared to long-lived GHGs (including CO_2 and CFCs). Our paper demonstrates the benefits of replacing high-GWP HFCs with low-GWP alternatives, so the overall forcing and associated warming due to HFC growth can be significantly reduced. The results presented here could strengthen the interest of policymakers in promoting fast-action strategies to reduce SLCPs, including HFCs, as a complement to immediate action to reduce CO_2 emissions. There are several policy options for limiting HFC growth, separate from those for BC and CH_4, including using the Montreal Protocol to phase down the production and consumption of HFCs (Molina et al., 2009; UNEP, 2012a, b), which would preserve the climate benefits the treaty has already achieved through its success in phasing out nearly 100 similar chemicals (Velders et al., 2007, 2012). Without the Montreal Protocol, the projected radiative forcing by ODSs would have been roughly $0.65 Wm^{-2}$ in 2010 (Velders et al., 2007), and the global temperature would have been higher (green line in Fig. 2). It is also important to emphasize that HFC mitigation should not be viewed as an "alternative" strategy for avoiding the 2 C warming, but rather as a critical component of a strategy that also requires mitigation of CO_2 and the other SLCPs. The focus of this study is on near-term warming over the next several decades to end of the century. For the longer-term (century and beyond), mitigation of CO_2 would be essential for a significant reduction in the warming.

REFERENCES

1. Andersen, S. O. and Morehouse, E. T.: The Ozone Challenge: Industry and Government Learned to Work Together To Protect Environment, American

Society of Heating, Refrigeration and Air-Conditioning Engineers (ASHRAE) Journal, 33–36, 1997.

2. Andersen, S. O. and Sarma, K. M.: Protecting the Ozone Layer: the United Nations History, Earthscan Press, London, 2002.

3. Andersen, S. O., Sarma, K. M., and Taddonio, K. N.: Technology Transfer for the Ozone Layer: Lessons for Climate Change, Earthscan Press, London, 2007.

4. Clarke, L., Edmonds, J., Jacoby, H., Pitcher, H., Reilly, J., and Richels, R.: Scenarios of Greenhouse Gas Emissions and Atmospheric Concentrations. Sub-report 2.1A of Synthesis and Assessment Product 2.1 by the US Climate Change Science Program and the Subcommittee on Global Change Research. Department of Energy, Office of Biological & Environmental Research, Washington, DC, USA, 154 pp., 2007.

5. Cofala J., Amann, M., Klimont, Z., Kupiainen, K., and Hoglund-Isaksson, L.: Scenarios of global anthropogenic emissions of air pollutants and methane until 2030, Atmos. Environ., 41, 8486– 8499, 2007.

6. EPA: Inventory of U.S. Greenhouse Gas Emissions and Sinks: 1990–2010, EPA 430-R-12-001, US Environmental Protection Agency, Washington DC, USA, 2012.

7. Forster, P. and Ramaswamy, V.: Changes in atmospheric constituents and in radiative forcing. Climate Change 2007: The Physical Sciences Basis. Contribution of Working Group I to the Fourth Assessment Report of the Intergovernmental Panel on Climate Change, edited by: Solomon, S., Qin, D., Manning, M., Chen, Z., Marquis, M., Averyt, K. B., Tignor, M., and Miller, H. L., 129–234, Cambridge Univ. Press, Cambridge, UK, 2007.

8. Gschrey, B., Schwarz, W., Elsner, C., and Engelhardt, R.: High increase of global F-gas emissions until 2050, Greenhouse Gas Measurement Management, 1, 85–92, 2011.

9. Hijioka, Y., Matsuoka, Y., Nishimoto, H., Masui, M., and Kainuma, M.: Global GHG emissions scenarios under GHG concentration stabilization targets. J. Global Environ. Eng., 13, 97–108, 2008.

10. IPCC: Special report on emissions scenarios, Intergovernmental Panel on Climate Change, Cambridge Univ. Press, Cambridge, UK and New York, 2000.

11. IEA: CO2 emissions from fuel combustion: Highlights, International Energy Agency, Paris, France, 2011.

12. Joos, F., Bruno, M., Fink, R., Siegenthaler, U., Stocker, T., Le Qu'er'e, C., and Sarmiento J.: An efficient and accurate representation of complex oceanic and biospheric models of anthropogenic carbon uptake, Tellus B Chem. Phys. Meteorol., 48, 397–417, 1996.

13. Meinshausen, M., Smith, S. J., Calvin, K., Daniel, J. S., Kainuma, M. L. T., Lamarque, J.-F., Matsumoto, K., Montzka, S. A., Raper, S. C. B., Riahi, K., Thomson, A., Velders, G. J. M., van Vuuren, D. P. P.: The RCP Greenhouse Gas Concentrations and their Extensions from 1765 to 2500, Clim. Change, 109, 213–241, doi:10.1007/s10584-011-0156-z, 2011.

14. Miller, B. R. and Kuijpers, L. J. M.: Projecting future HFC-23 emissions, Atmos. Chem. Phys., 11, 13259–13267, doi:10.5194/acp- 11-13259-2011, 2011.

15. Molina, M. and Rowland, F. S.: Stratospheric Sink for Chlorofluoromethanes: Chlorine Atom-Catalyzed Destruction of Ozone, Nature, 249, 810–814, 1974.
16. Molina M., Zaelke, D., Sarma, K. M., Andersen, S. O., Ramanathan, V., and Kaniaru, D.: Reducing abrupt climate change risk using the Montreal Protocol and other regulatory actions to complement cuts in CO2 emissions, Proc. Natl. Acad. Sci., 106, 20616–20621, 2009.
17. Pinnock, S., Hurley, M. D., Shine, K. P., Wallington, T. J., and Smyth, T. J.: Radiative forcing of climate by hydrochlorofluorocarbons and hydrofluorocarbons, J. Geophys. Res., 100, 23227–23238, 1995.
18. Ramanathan V.: Greenhouse Effect Due to Chlorofluorocarbons: Climatic Implications, Science, 190, 50–52, 1975.
19. Ramanathan V. and Xu, Y.: The Copenhagen Accord for limiting global warming: Criteria, constraints, and available avenues, Proc. Natl. Acad. Sci., 107, 8055–8062, 2010.
20. Ramanathan, V., Cicerone, R. J., Singh, H. B., and Kiehl, J. T.: Trace gas trends and their potential role in climate change, J. Geophys. Res., 90, 5547–5566, 1985a.
21. Ramanathan, V., Callis, L., Cess, R., Hansen, J., Isaksen, I., Kuhn, W., Lacis, A., Luther, F., Mahlman, J., Reck, R., and Schlesinger, M.: Trace Gas Effects on Climate, Assessment of our understanding of the processes controlling its present distribution and change, WMO, Global Ozone Research and Monitoring Project, Report, no. 16. Report commissioned by NASA/Federal Aviation Administration, NOAA, WMO, UNEP, Commission of the European Communities and Bundesminisiterium Fur Forschung Technologie, Chapter 16, Volume III of Atmospheric Ozone 1985, 821–863, 1985b.
22. Ravishankara, A. R., Turnipseed, A. A., Jensen, N. R., Barone, S., Mills, M., Howard, C. J., and Solomon, S.: Do hydrocarbons destroy stratospheric ozone?, Science, 263, 71–75, 1994.
23. Riahi, K., Gruebler, A., and Nakicenovic, N.: Scenarios of longterm socio-economic and environmental development under climate stabilization, Technol. Forecast. Soc. Change, 74, 887–935 2007.
24. Royal Society: Ground-level ozone in the 21st century: Future trends, impacts and policy implications, The Royal Society, London, UK, 23–54, 2008.
25. Shindell, D., Kuylenstierna, J. C. I., Vignati, E., van Dingenen, R., Amann, M., Klimont, Z., Anenberg, S. C., Muller, N., Janssens- Maenhout, G., Raes, F., Schwartz, J., Faluvegi, G., Pozzoli, L., Kupiainen, K., H"oglund-Isaksson, L., Emberson, L., Streets, D., Ramanathan, V., Hicks, K., Oanh, N. T. K., Milly, G., Williams, M., Demkine, V., and Fowler, D.: Simultaneously Mitigating Near-Term Climate Change and Improving Human Health and Food Security, Science, 335, 183–189, 2012.
26. Smith, S. J., West, J. J., and Kyle, P.: Economically Consistent Long-Term Scenarios for Air Pollutant and Greenhouse Gas Emissions, Clim. Change, 108, 619–627, 2011.
27. Smith, S. M., Lowe, J. A., Bowerman, N. H. A., Gohar, L. K., Huntingford, C., and Allen, M. R.: Equivalence of greenhouse-gas emissions for peak temperature limits, Nature Clim. Change 2, 535–538, 2012.

28. Thomson A. M., Calvin, K. V., Smith, S. J., Kyle, G. P., Volke, A. C., Patel, P. L., Delgado Arias, S., Bond-Lamberty, B., Wise, M. A., Clarke, L. E., and Edmonds, J. A.: RCP4.5: A Pathway for Stabilization of Radiative Forcing by 2100, Clim. Change, 109, 77–94, doi:10.1007/s10584-011-0151-4, 2011.

29. UNEP: Adjustments agreed by the Nineteenth Meeting of the Parties relating to the controlled substances in group I of Annex C of the Montreal Protocol (hydrochlorofluorocarbons), in Report of the Nineteenth Meeting of the Parties to the Montreal Protocol on Substances that Deplete the Ozone Layer, UNEP/OzL.Pro.19/7, United Nations Environment Program Ozone Secretariat, Nairobi, Kenya, 2007.

30. UNEP: Report of the Twenty-Second Meeting of the Parties to the Montreal Protocol on Substances that Deplete the Ozone Layer, UNEP/Ozl.Pro.22/9, United Nations Environment Program Ozone Secretariat, Nairobi, Kenya, 2010.

31. UNEP: HFCs: A Critical Link In Protecting Climate and the Ozone Layer, United Nations Environment Programme, Nairobi, Kenya, 2011.

32. UNEP: Proposed amendment to the Montreal Protocol submitted by the Federated States of Micronesia, UNEP/OzL/Pro.WG.1/32/5, United Nations Environment Program, Nairobi, Kenya. 2012a.

33. UNEP: Proposed amendment to the Montreal Protocol submitted by Canada, Mexico and the United States of America, UNEP/OzL.Pro.WG.1/32/6, United Nations Environment Program, Nairobi, Kenya, 2012b.

34. UNEP and WMO: Integrated Assessment of Black Carbon and Tropospheric Ozone, United Nations Environment Program and World Meteorological Organization, Nairobi, Kenya, 2011.

35. van Vuuren, D., den Elzen, M., Lucas, P., Eickhout, B., Strengers, B., van Ruijven, B., Wonink, S., and van Houdt, R.: Stabilizing greenhouse gas concentrations at low levels: an assessment of reduction strategies and costs, Clim. Change, 81, 119–159, doi:10.1007/s/10584-006-9172-9, 2007.

36. van Vuuren, D. P., Edmonds, J., Kainuma, M., Riahi, K., Thomson, A., Hibbard, K., Hurtt, G. C., Kram, T., Krey, V., Lamarque, J.- F., Masui, T., Meinshausen, M., Nakicenovic, N., Smith, S. J., and Rose, S. K.: The Representative Concentration Pathways: An Overview, Clim. Change, 109, 5–31, 2011.

37. Velders, G. J. M., Andersen, S. O., Daniel, J. S., Fahey, D. W., and McFarland, M.: The Importance of the Montreal Protocol in Protecting the Climate, Proc. Natl. Acad. of Sci., 104, 4814–4819, 2007.

38. Velders, G. J. M., Fahey, D. W., Daniel, J. S., McFarland, M., and Andersen, S. O.: The Large Contribution of Projected HFC Emissions to Future Climate Forcing, Proc. Natl. Acad. of Sci., 106, 10949–10954, 2009.

39. Velders, G. J. M., Ravishankara, A. R., Miller, M. K., Molina, M. J., Alcamo, J., Daniel, J. S., Fahey, D. W., Montzka, S. A., and Reimann, S.: Preserving Montreal Protocol Climate Benefits by Limiting HFCs, Science, 335, 922–923, 2012.

40. Wallack, J. and Ramanathan, V.: The other climate changers, why black carbon also matters, Foreign Aff., 88, 105–113, 2009.

41. Wang, W. C., Yung, Y. L., Lacis, A. A., Mo, T., and Hansen, J. E.: Greenhouse effect due to manmade perturbations of trace gases, Science, 194, 685–690, 1976.

42. Wise, M. A., Calvin, K. V., Thomson, A. M., Clarke, L. E., Bond- Lamberty, B., Sands, R. D., Smith, S. J., Janetos, A. C., and Edmonds, J. A.: Implications of Limiting CO2 Concentrations for Land Use and Energy, Science, 324, 1183–1186, 2009.
43. WMO: Report No. 52: Scientific Assessment of Ozone Depletion: 2010, World Meteorological Organization Global Ozone Research and Monitoring Project, Geneva, Switzerland, available at http://ozone.unep.org/Assessment Panels/SAP/ Scientific Assessment 2010/00-SAP-2010-Assement-report.pdf, 2011.

CHAPTER 6

Analysis of Thermodynamic Characteristic Changes in Direct Expansion Ground Source Heat Pump Using Hydrofluoroolefins (HFOs) Substituting for HFC-134a

YUEFEN GAO, HONGLEI ZHAO, YINGXIN PENG, AND TONY ROSKILLY

6.1 INTRODUCTION

Direct expansion ground source heat pump (DXGSHP) has only copper loops circulating refrigerants which exchanges heat directly with the soil through the walls of the copper tubing. It is an energy-efficient and environmentally clean space conditioning system. However, it has the risk of ground contamination if refrigerant leak into the ground, as the loops containing refrigerant are directly buried in the ground.

One of the solutions to this problem is to select non-toxic, environment-amiable working fluids. Generally, commercial and residential heat pump systems use hydrofluorocarbons (HFCs) or their mixtures as refrigerants. HFCs have zero ozone depletion potential (ODP), but most of them have relatively high values of global warming potentials (GWP). In developed

© 2013 Yuefen Gao et al. Energy and Power Engineering, *Vol. 5 No. 2A (2013)*, *Article ID: 30599*, *DOI:10.4236/epe.2013.52A002. Creative Commons Attribution License (http://creativecommons.org/licenses/by/3.0/).*

countries, refrigerants with high GWP are facing to be phased out. The European Union's F-gas regulation and the directive 2006/40/EC ban fluorinated gases having GWP greater than 150 in new mobile models from January 1, 2011 and in new vehicles from January 1, 2017 [1,2]. More efforts have been underway to investigate fluorinated propene isomers as possible refrigerants. In 2007, DuPont and Honeywell co-developed hydrofluoroolefins (HFOs) to replace HFCs in air conditioning units. HFOs are a class of compounds. Among the compounds, HFO-1234yf and HFO-1234ze[E] are the two most suitable to the air conditioning system [3].

HFO-1234yf and HFO-1234ze[E] possess the similar thermodynamic behavior to HFC-134a. They have a 100 year GWP of 4 and 6 respectively. Both of them are being considered as a possible replacement for HFC-134a which has a high GWP value of 1430 [3]. The US EPA has released a proposed rule for HFO-1234yf as an automotive refrigerant [4].

Not only are HFOs able to be used to the air conditioning system in vehicles, they are also feasible to replace HFC-134a in commercial and residential heat pump system. Many investigations released that HFO-1234yf and HFO-1234ze[E] almost have non-toxic and environmental impact [3,5-10]. They are also compatible with the lubricate oil and materials normally used in HFC-134a system [5,11]. HFO-1234ze[E] is non-flammable, while HFO-1234yf is quite mildly flammable [3]. So they are ideal options for DXGSHP in terms of climate change and safety.

Some research results also indicated that HFOs' thermophysical property parameters are to some extent different from HFC-134a's [3,5,12-15]. Correspondingly, the DXGSHP system possibly has different thermodynamic performance with these refrigerants.

This study aims to compare the thermodynamic performances of the DXGSHP system using the three kinds of refrigerants and to recommend the suitable alternative substance for HFC-134a based on theoretical calculation and analysis.

6.2 DXGSHP SYSTEM DESCRIPTION

In the DXGSHP system, the refrigerant loops are directly buried underground and exchange heat with the ground.

Without using an intermediate fluid, the DXGSHP system is significantly more efficient than the conventional GSHP system. Above all, the phase change of the refrigerant occurs in the ground, where latent heat is efficiently rejected or absorbed. The elimination of the water pump and the water heat exchanger also greatly reduce electricity consumption and heat losses.

Figure 1 shows a schematic diagram of DXGSHP system. The system operates in both heating and cooling mode. A reversal valve transforms the operating mode.

In the heating mode, refrigerant absorbs heat from the ground through the loops named ground heat exchanger (GHX). It then enters the compressor where it is compressed to high-temperature vapor. The discharged vapor then enters air heat exchanger (AHX) where it releases heat to the space and condenses to liquid. The liquid exits from the AHX and is throttled via the expansion valve (EV). It then enters the GHX.

In the cooling mode, inversely, the refrigerant releases heat to the ground via GHX. The refrigerant absorbs heat from the space via AHX. It then enters the compressor and is compressed. The discharged vapor enters GHX and condenses to liquid. The liquid exits from the GHX and is throttled via the EV, and finally enters the AHX.

The thermodynamic cycle is expressed in the p-h diagram (Figure 2). Both the modes have the same cycle diagram except for the parameters.

6.3 SIMULATING CALCULATION OF THE THERMODYNAMIC PERFORMANCE OF DXGSHP

In order to compare the thermodynamic performance of the DXGSHP using the three different kinds of refrigerants, simulating calculation are conducted based on the following conditions and assumptions. In the thermodynamic calculations, the thermophysical property parameters of each state point are based on the program REF PROP8.0 for HFC-134a, Extended Corresponding States (ECS) model for HFO-1234ze(E) [16] and Martin-Hou Equation of State (EOS) for HFO-1234yf [5]. The ECS model has adequate accuracy [16,17].

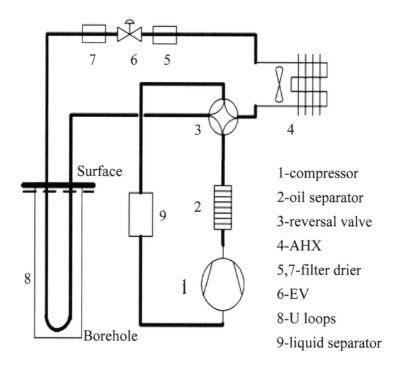

Figure 1.The schematic of the DX-GCHP system.

6.3.1 PARAMETERS AT DESIGN CONDITIONS

Evaporating temperature and condensing temperature are the two most important parameters in the system design. The temperatures depend much on the external parameters. According to ANSI-AHRI870-2005, in the cooling mode, air temperature entering indoor AHX is 26.7°C dry-bulb, 19.4°C wet-bulb. Refrigerant temperature of liquid line from GHX is 25.0°C. In the heating mode, air temperature entering AHX is 21.1°C dry-bulb, 15.6°C wet-bulb. Refrigerant temperature of vapor line from

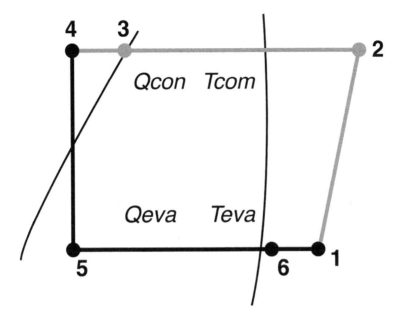

Figure 2. The p-h diagram of thermodynamic cycle.

GHX is 0°C [18]. The values of evaporating temperature and condensing temperature are presented in Table 1.

6.3.2 ASSUMPTIONS IN THE THERMODYNAMIC CALCULATION

- All the processes are under steady state conditions. • There are no potential or kinetic energy effects and no chemical or nuclear reactions.
- Heat losses and refrigerant pressure drops in the connecting tubes are negligible.
- The isentropic efficiency of the compressor is 85%. The compressor mechanical efficiency and the compressor motor electrical efficiency are 70% and 75%, respectively.

Table 1. Parameters at design conditions.

Parameters	Cooling mode	Heating mode
Air temperature entering AHX	26.7°C dry-bulb 19.4°C wet-bulb	21.1°C dry-bulb 15.6°C wet-bulb
Evaporating temperature, teva	7°C	0°C
Superheat degree, Δtsh	5°C	5°C
Suction temperature, tsuc	17°C	10°C
Condensing temperature, tcon	30°C	35°C
Subcooling degree, Δtsc	5°C	5°C

• The systems with different refrigerants have the same compressing speed.

6.3.3 MODELS ASSESSING THE THERMODYNAMIC PERFORMANCE OF DXGSHP SYSTEM

The performance of the system is evaluated primarily in terms of the capacity and coefficient of performance (COP). The parameters are calculated according to the steady flow energy equations.

The capacity is related to the operation parameters and the refrigerant mass flow rate. The relations are expressed as

$$Q_{eva} = m_r(h_{eva,o} - h_{eva,i}) \tag{1}$$

$$Q_{con} = m_r(h_{con,i} - h_{con,o}) \tag{2}$$

$$m_r = \eta_v \frac{V_h}{VS_{uc}} \tag{3}$$

$$\eta_v = A - B\left(\frac{P_{dis}}{P_{suc}}\right)^{\frac{1}{n}} \tag{4}$$

where, η_v is the volumetric efficiency, A and B are the coefficients related to the types of the compressor, n is the specific heat ratio related to the refrigerant type.

The power input is related to the operation parameters, the mass flow rate, the mechanical efficiency of the compressor and the electrical efficiency of the driving motor.

$$W_{com} = m_r \frac{h_{dis} - h_{suc}}{\eta_m \eta_e} \qquad (5)$$

η_m tends to decrease with the increase of pressure ratio.

In the heating mode COP_h is the ratio of the heating capacity Q_{con} to the power input W_{com}, whereas, in the cooling mode COP_c is the ratio of the cooling capacity $Q_{eva,m}$ to the power input W_{com}, as given below respectively.

$$COP_h = \frac{Q_{con}}{W_{com}} \qquad (6)$$

$$COP_c = \frac{Q_{eva}}{W_{com}} \qquad (7)$$

Combining the equations of (1), (5), (7) and (2), (5), (6), COP_h and COP_c can be expressed as

$$COP_h = \eta_m \eta_e \frac{h_{con,i} - h_{con,o}}{h_{dis} - h_{suc}} \qquad (8)$$

$$COP_c = \eta_m \eta_e \frac{h_{eva,o} - h_{eva,i}}{h_{dis} - h_{suc}} \qquad (9)$$

6.4 RESULTS AND ANALYSIS

6.4.1 CALCULATING RESULTS

Based on the parameters in Table 1 and the assumptions above, the calculating results are shown in Tables 2-4.

Table 2 states the typical characteristic parameters and performance values. In the theoretical cycle, HFC-134a has the highest capacity per unit mass flow rate. HFO-1234ze[E] has the closest values to HFC-134a. However, HFO-1234yf consumes the lowest power among them. The ratio of $COP_{th,h}$ of the three kinds of refrigerants (HCF-134a, HFO-1234yf, HFO-1234ze[E]) is 1:0.995: 1.041, while the ratio of $COP_{th,c}$ of the three kinds of refrigerants is 1:1.085:1.141. That means HFO-1234ze[E] has the best thermodynamic performance both in the heating mode and in the cooling mode, while HFO- 1234yf has the lowest performance in the heating mode. But, in the actual cycle, the ratio of COP_h of the three kinds of refrigerants (HCF-134a, HFO-1234yf, HFO-1234ze[E]) is 1:1.028:0.955, while the ratio of COP_c is 1:1.121:1.008. HFO-1234yf has the best thermodynamic performance.

Table 3 shows the calculating values and the ratios under the conditions that all refrigerants have the same cooling capacity and the mass flow rate of each refrigerant itself keeps the same in each operating mode. The system using HFO-1234yf needs 23% more mass charge than the system using HFC-134a, while the mass charge of the system using HFO-1234ze[E] is nearly 10% more than that of the system using HFC-134a. Under the same condition, the volume flow rate of HFO-1234ze[E] system is 24% - 26% more than that of HFC-134a system.

Table 4 gives the results that all refrigerants have the same volume flow rate in each operating mode. Refrigerant charge is apt for the maximum in each operating mode. The mass flow rate is regulated according to the operating mode. A reservoir is installed to contain the redundant refrigerant. The results show that refrigerant charge of HFO-1234yf system is 21% more than that of HFC-134a system, whereas cooling capacity and heating capacity of HFO-1234yf system are 2% and 8% less than those of HFC-134a system respectively. As for HFO-1234ze[E] system, refrigerant charge of the system is 14% less than that of HFC-134a system, whereas heating capacity and cooling capacity of HFO-1234ze[E] system are about 20% less than those of HFC-134a system.

6.4.2 ANALYSES AND DISCUSSIONS

6.4.2.1 THERMOPHYSICAL PROPERTIES

1) Pressure ratio: As for the refrigerating cycle, the ratio of the discharge pressure p_{dis} to the suction pressure p_{suc} affects the thermodynamic performance significantly. Equations (1)-(9) show that the volumetric

Table 2. Thermodynamic parameters.

Variables	Cooling mode			Heating mode		
	HFC-134a	HFO-1234yf	HFO-1234ze[E]	HFC-134a	HFO-1234yf	HFO-1234ze[E]
v_{suc}, m³/kg	0.058	0.048	0.067	0.073	0.063	0.083
p_{dis}/p_{suc}	2.03	1.95	2.15	2.98	2.86	3.05
t_{dis}, °C	43	38	40	48	44	44
$q_{eva,v}$, kJ/m³	2924.8	2864.6	2324.9	2154.4	1989.4	1742.3
$q_{eva,m}$, kJ/kg	169.639	137.500	155.769	157.273	125.334	144.615
q_{con}, kJ/kg	186.363	150.000	169.231	181.818	145.000	166.153
w_{com}, kJ/kg	27.878	20.161	25.400	40.908	31.721	39.160
$COP_{th,h}$	-	-	-	7.407	7.373	7.714
$COP_{th,c}$	10.142	11.000	11.571	-	-	-
COP_h	-	-	-	4.445	4.571	4.243
COP_c	6.085	6.820	6.133	-	-	-

efficiency η_v, the mass flow rate m_r, the cooling capacity Q_{eva} or heating capacity Qcon, the power consumption W_{com} and COP are the functions of the pressure ratio. The adiabatic indicated efficiency and the friction efficiency decline as the pressure ratio increases.

HFO-1234ze[E] has the highest pressure ratio (p_{dis}/p_{suc}), while HFO-1234yf has the lowest. In Table 2, p_{dis}/p_{suc} of HFO-1234ze[E] and HFO-1234yf are 2.15 and 1.95 respectively in the cooling mode, 3.05 and 2.86 in the heating mode. So, the higher pressure ratio of HFO-1234ze[E] in an actual cycle offsets its original advantages. The ratio of COPh of the three kinds of refrigerants (HCF-134a, HFO-1234yf, HFO-1234ze[E]) is 1:1.028:0.955, while the ratio of COPc is 1:1.121:1.008. As discussed above, in the theoretical cycle, HFO-1234ze[E] has the highest value of COP both in the cooling and heating mode. However, in the actual cycle HFO-1234yf obviously has the best thermodynamic performance because of its lower pressure ratio.

Table 3. Thermodynamic variables based on the same cooling capacity.

Variables	Cooling mode			Heating mode		
	HFC-134a	HFO-1234yf	HFO-1234ze[E]	HFC-134a	HFO-1234yf	HFO-1234ze[E]
$V_r \times 10^{-3}$, m³/s	2.051	2.094	2.581	2.583	2.747	3.195
m_r, kg/s	0.0354	0.0436	0.0385	0.0354	0.0436	0.0385
Q_{eva}, kW	6	6	6	5.563	5.465	5.568
Q_{con}, kW	6.592	6.545	6.518	6.431	6.319	6.395
W_{com}, kW	0.987	0.879	0.978	1.447	1.383	1.508
$m_{r1}:m_{r2}:m_{r3}$		1:1.23:1.09			1:1.23:1.09	
$V_{r1}:V_{r2}:V_{r3}$		1:1.02:1.26			1:1.06:1.24	
$Q_{con1}:Q_{con2}:Q_{con3}$		1:0.993:0.989			1:0.983:0.994	
$Q_{eva1}:Q_{eva2}:Q_{eva3}$		1:1:1			1:0.982:1.001	

2) Working pressure and temperature: When the systems are running under the same conditions, the suction pressure and the discharge pressure of HFO-1234yf are slightly higher than those of HCF-134a. However, HFO-1234ze[E] is just the reverse. The discharge vapor temperature of both HFO-1234yf and HFO-1234ze[E] is lower than HCF-134a. So when HFO-1234yf or HFO-1234ze[E] directly substitutes for HFC-134a, the compressor or the system still works in safety, as the maximum working pressure and temperature are kept below or close to the values of the HFC-134a system.

3) Suction specific volume v_{suc}: Suction specific volume v_{suc} affects the volume flow, and further the size of the compression chamber, evaporator, condenser and pipes. HFO-1234ze[E] has the largest specific volumes (0.067 m³/kg in the cooling mode, 0.083 m³/kg in the heating mode), while HFO-1234yf has the smallest ones (0.048 m³/kg in the cooling mode, 0.063 m³/kg in the heating mode). So, the capacity per unit of swept volume in the HFO-1234ze[E] system is the smallest. Larger size compression chamber, evaporator, condenser and pipes are needed. This will raise the initial investment.

6.4.2.2 REFRIGERANT CHARGE

1) Refrigerant mass charge varying in different systems: Two cases are discussed here. The first case is that the three systems provide the same cooling capacity. The second case is that the volume flow rates in the systems keep at a given value.

In the first case, more refrigerant charge is required in the HFO-1234yf and HFO-1234ze[E] systems when they provide the same cooling capacity. The data in Table 3 show that the system using HFO-1234yf needs 23% more mass charge than the system using HFC-134a, while the mass charge of the system using HFO-1234ze[E] is nearly 10% more than that of the system using HFC-134a.

According to the results in Table 3, the volume flow rate of HFO-1234ze[E] system is 24-26% more than that of HFC-134a system. So the compression chamber and the pipes of HFO-1234ze[E] system should have larger sizes. The capacities of condenser in the heating mode (i.e.,

Table 4. Thermodynamic variables with the same volume flow rate.

Variables	Cooling mode			Heating mode		
	HFC-134a	HFO-1234yf	HFO-1234ze[E]	HFC-134a	HFO-1234yf	HFO-1234ze[E]
$V_r \times 10^{-3}$, m³/s	2.051	2.051	2.051	2.051	2.051	2.051
m_r, kg/s	0.0354	0.0427	0.0306	0.0281	0.0326	0.247
Q_{eva}, kW	6	5.871	4.773	4.418	4.081	3.574
Q_{con}, kW	6.592	6.404	5.186	5.107	4.721	4.106
W_{com}, kW	0.987	0.861	0.778	1.149	1.034	0.968
$m_{r1}:m_{r2}:m_{r3}$		1:1.21:0.86			1:1.16:0.88	
$V_{r1}:V_{r2}:V_{r3}$		1:1:1			1:1:1	
$Q_{con1}:Q_{con2}:Q_{con3}$		1:0.971:0.787			1:0.924:0.804	
$Q_{eva1}:Q_{eva2}:Q_{eva3}$		1:0.979:0.796			1:0.924:0.809	

the heating capacities) of HFO-1234yf and HFO-1234ze[E] system are 2% and 1% less than that of HFC-134a system respectively.

In the second case, when HFC-134a is replaced with HFO-1234yf or HFO-1234ze[E] in a given unit, the heating capacity and the cooling capacity will decrease.

The results in Table 4 show that refrigerant charge of HFO-1234yf system is 21% more than that of HFC-134a system, whereas cooling capacity and heating capacity of HFO-1234yf system are 2% and 8% less than those of HFC-134a system respectively. As for HFO-1234ze[E] system, refrigerant charge of the system is 14% less than that of HFC-134a system, whereas heating capacity and cooling capacity of HFO-1234ze[E] system are about 20% less than those of HFC-134a system.

2) More refrigerant charge required in the cooling mode: More refrigerant charge is required in the cooling mode than in the heating mode.

The results in Table 4 show that 24% - 31% more refrigerant charge is required in the cooling mode than in the heating mode.

6.4.2.3 COMPRESSION CHAMBER VOLUME

HFO-1234yf and HFO-1234ze[E] need larger compression chamber volume if they provide the same heating or cooling loads, because HFO-1234yf and HFO-1234ze[E] have the smaller capacity per unit of swept volume $q_{eva,v}$ than HFC-134a. According to Table 2, $q_{eva,v}$ of HFC-134a, HFO-1234yf and HFO-1234ze[E] systems are 2924.8 kJ/m³, 2864.6 kJ/m³, 2324.9 kJ/m³ respectively in the cooling mode, and 2154.4 kJ/m³, 1989.4 kJ/m³, 1742.3 kJ/m³ in the heating mode.

6.4.2.4 CAPACITY AND COP

Capacity and COP characterize the capability and performance of the system.

As for the cooling or heating capacity, HFC-134a system has the largest capacity per unit of swept volume qeva,v. HFO-1234yf system has the similar value to HFC-134a system. But the capacity of HFO-1234ze[E] system is about 20% less than that of HFC-134a system. That means HFO-1234yf is the better replacement for HFC-134a than HFO-1234ze[E].

In terms of COP, HFO-1234yf has the better performance than the other two in the actual cycle, as discussed above.

6.5 CONCLUSIONS

In the DXGSHP system, the refrigerant loops are directly buried in the ground, which is in risk of ground contamination. Using non-toxic refrigerants is one effective way to solve this problem. Both HFO-1234yf and HFO-1234ze[E] are ideal potential substituent for HFC-134a in terms of climate change and safety.

HFO-1234ze[E] has the best thermodynamic performance assumed that all the refrigerants have the same mass flow rate. However, in an actual cycle HFO-1234yf has the best thermodynamic performance due to its low pressure ratio.

More refrigerant mass charges are required in the HFO-1234yf and HFO-1234ze[E] system when keeping the heating capacity the same. Moreover, more refrigerant mass charge is required in the cooling mode than in the heating mode.

HFO-1234yf and HFO-1234ze[E] have so smaller capacity per unit of swept volume that they need larger chamber if providing the same heating or cooling loads. For a given unit when HFC-134a is replaced with HFO-1234yf or HFO-1234ze[E], the capacity will decrease.

In conclusion, HFO-1234yf is the better potential alternative substance for HFC-134a. HFO-1234ze[E] can also substitute for HFC-134a, but it needs larger size unit.

REFERENCES

1. The European Parliament and the Council of the European Union, "Regulations (EC) No 842/2006 of the European Parliament and of the Council of 17 May 2006 on Certain Fluorinated Greenhouse Gases," Official Journal of the European Union, 2006. http://eur-lex.europa.eu/LexUriServ/LexUriServ.do?uri=OJ:L:2006:161:0001:0011:EN:PDF

2. The European Parliament and the Council of the European Union, "Directive 2006/40/EC of the European Parliament and of the Council of 17 May 2006 Relating to Emissions from Air-Conditioning Systems in Motor Vehicles and Amending Council Directive 70/156/EEC," Official Journal of the European Union, 2006. http://eur-lex.europa.eu/LexUriServ/LexUriServ.do?uri=OJ:L:2006 :161:0012:0018:EN:PDF

3. T. G. A. Vink, "Synthetic and Natural Refrigeration Fluids Recent Developments," 2012. http://www.fluorocarbons.org/uploads/Modules/Library/tvink_honeywell_ecocool10.pdf

4. USEPA, "Protection of Stratospheric Ozone: New Substitute in the Motor Vehicle Air Conditioning Sector Under the Significant New Alternatives Policy (SNAP) Program," Federal Register, Vol. 74, No. 200, 2009. http://www.gpo.gov/fdsys/pkg/FR-2009-10-19/pdf/E9-25106.pdf

5. T. J. Leck, "Evaluation of HFO-1234yf as a Potential Replacement for R-134a in Refrigeration Applications," Third IIR Conference on Thermophysical Properties and Transfer Processes of Refrigerants, Boulder, 23-26 June 2009. http://www2.dupont.com/Refrigerants/en.../HFO-1234yf_IIR_Leck.pdf

6. C. Zilio, J. S. Brown, G. Schiochet and A. Cavallini, "The Refrigerant R1234yf in Air Conditioning Systems," Energy, 36, No. 10, 2011, pp. 6110-6120. doi:10.1016/j.energy.2011.08.002

7. P. Schuster, R. Bertermann, G. M. Rusch and W. Dekant, "Biotransformation of trans-1,1,1,3-tetrafluoropropene (HFO-1234ze)," Toxicology and Applied Pharmacology Vol. 239, No. 3, 2009, pp. 215-223. doi:10.1016/j.taap.2009.06.018

8. P. Schuster, R. Bertermann, T. A. Snow, X. Han, G. M. Rusch and G. W. Jepson, "Biotransformation of 2,3,3,3- tetrafluoropropene (HFO-1234yf)," Toxicology and Applied Pharmacology, Vol. 233, No. 3, 2008, pp. 323-332. doi:10.1016/j.taap.2008.08.018

9. T. J. Wallington, M. P. S. Andersen and O. J. Nielsen, "Estimated Photochemical Ozone Creation Potentials (POCPs) of CF3CFCH2 (HFO-1234yf) and Related Hydrofluoroolefins (HFOs)," Atmospheric Environment, Vol. 44, No. 11, 2010, pp. 1478-1481. doi:10.1016/j.atmosenv.2010.01.040

10. K. Takizawa, K. Tokuhashi and S. Kondo, "Flammability Assessment of CH2CFCF3: Comparison with Fluoroalkenes and Fluoroalkanes," Journal of Hazardous Materials, Vol. 172, No. 2-3, 2009, pp. 1329-1338. doi:10.1016/j.jhazmat.2009.08.001

11. B. Wang, W. Zhang and J. Lv, "A New Refrigerant HFO- 1234ze," New Chemical Materials, Vol. 36, No. 2, 2008, pp. 10-12. (in Chinese)

12. K. Tanaka and Y. Higashi, "Thermodynamic Properties of HFO-1234yf (2,3,3,3-tetrafluoropropene)," International Journal of Refrigeration, Vol. 33, No. 3, 2010, pp. 474-479. doi:10.1016/j.ijrefrig.2009.10.003

13. Y. Higashi and K. Tanaka, "Critical Parameters and Saturated Densities in the Critical Region for trans-1,3,3,3- Tetrafluoropropene (HFO-1234ze(E))," Journal of Chemical & Engineering Data, Vol. 55, No. 4, 2010, pp. 1594-1597. doi:10.1021/je900696z

14. USEPA, "Transitioning to Low-GWP Alternatives in Unitary Air Conditioning," US Environmental Protection Agency, 2010. http://www.epa.gov/ozone/downloads/EPA_HFC_UAC.pdf

15. D. Del Col, D. Torresin and A. Cavallini, "Heat Transfer and Pressure Drop during Condensation of the Low GWP Refrigerant R1234yf," International Journal of Refrigeration, Vol. 33, No. 7, 2010, pp. 1307-1318. doi:10.1016/j.ijrefrig.2010.07.020

16. R. Akasaka, "An Application of the Extended Corresponding States Model to Thermodynamic Property Calculations for trans-1,3,3,3-tetrafluoropropene (HFO-1234 ze(E))," International Journal of Refrigeration, Vol. 33, No. 5, 2010, pp. 907-914. doi:10.1016/j.ijrefrig.2010.03.003

17. R. Akasaka, K. Tanaka and Y. Higashi, "Thermodynamic Property Modeling for 2,3,3,3-tetrafluoropropene (HFO- 1234yf)," International Journal of Refrigeration, Vol. 33, No. 1, 2010, pp. 52-60. doi:10.1016/j.ijrefrig.2009.09.004

18. AHRI, "ANSI/AHRI Standard 870-2005 with Addendum 1, Performance Rating of Direct GeoExchange Heat Pumps," Air-Conditioning, Heating, and Refrigeration Institute, Arlington, 2009.

CHAPTER 7

Inventories and Scenarios of Nitrous Oxide Emissions

ERIC A. DAVIDSON AND DAVID KANTER

7.1 INTRODUCTION

As the third most important anthropogenic greenhouse gas and the largest remaining anthropogenic stratospheric ozone depleting substance currently emitted, nitrous oxide (N_2O) is one of the most important forms of nitrogen (N) pollution (Ravishankara et al 2009, Ciais et al 2013). Excess N pollution has been identified as one of the three global environmental issues whose 'planetary boundary' has been surpassed (Rockström et al 2009). Once an N atom is in a reactive form, it can contribute to a number of cascading environmental problems as it is transported through terrestrial and aquatic ecosystems and into the atmosphere (Galloway et al 2003). Effective mitigation for N_2O emissions requires understanding of the sources and how they may increase this century.

N_2O is a by-product of several fundamental natural reactions of the N cycle: nitrification, denitrification, and chemo-denitrification

© 2014 IOP Publishing Ltd., Eric A Davidson and David Kanter. 2014 Environ. Res. Lett. 9 105012. DOI: http://dx.doi.org/10.1088/1748-9326/9/10/105012.

(Firestone and Davidson 1989). Humans began altering the natural N cycle as they expanded agricultural land, used fire as a land clearing and management tool, and cultivated leguminous crops that carry out biological N fixation. Human alteration accelerated dramatically with the discovery of the Haber–Bosch process, the chemical process that synthetically transforms atmospheric dinitrogen (N_2) gas into ammonia (NH_3) (Erisman et al 2008). The industrial production of NH_3 led to the development synthetic N fertilizers, which play a central role in feeding the world's rapidly increasing population. Without the Haber–Bosch process, about half of the world's population today would likely not be adequately nourished (Erisman et al 2008). This growth in anthropogenically-fixed N has simultaneously led to an unintended increase in global N pollution, including N_2O emissions, driven largely by the fact that mismatch between crop N demand and soil N supply frequently leads to N losses. With the possible exception of certain industrial point sources, it is impossible to completely eliminate global N pollution, particularly from agriculture—its largest source.

This paper first updates constraints on estimates and their uncertainties for anthropogenic and natural components of N_2O emissions by biome and by anthropogenic sector. We then consider a suite of emissions scenarios for N_2O, including those of the recent Representative Concentration Pathways (RCPs) of the Intergovernmental Panel on Climate Change (IPCC, van Vuuren et al 2011a) and a recent special United Nations report on N_2O (UNEP 2013), and the magnitude of mitigation efforts that would be needed to stabilize atmospheric N_2O by 2050. Future potential emissions from biofuels are discussed separately given particularly high levels of uncertainty for this sector. This paper integrates the authors' contributions to that UNEP report, other chapters in that report, the IPCC fifth assessment (AR5), and other recent literature for the most recent information on N_2O.

7.2 NATURAL EMISSIONS

The first approach to emission estimation is called 'bottom-up,' because it sums up emission inventories from field measurements, organized according to ecosystem type or by geographic region. Using the 'bottom-up' approach, published central estimates of current natural emissions of N_2O from terrestrial, marine and atmospheric sources based on several inventories range from 10 to 12 Tg N_2O-N yr^{-1} (Mosier et al 1998, Galloway et al 2004, Crutzen et al 2008, Syakila and Kroeze 2011). The IPCC AR5 (Ciais et al 2013) estimated that current natural sources of N_2O add up to roughly 11 (range 5.4–19.6) Tg N_2O-N yr^{-1}, which is the sum of emissions from terrestrial (6.6; range 3.3–9.0), marine (3.8; range 1.8–9.4) and atmospheric sources (0.6; range: 0.3–1.2; see figure 1). Note that the indicated uncertainty ranges from each bottom-up estimate are added together to produce the large range about the AR5 global estimate. Combining estimates of natural and anthropogenic emissions from Ciais et al 2013 and the 2013 UNEP report (see section 3) respectively, we estimate that natural emissions account for approximately two thirds of total global N_2O emissions (figure 1).

A second approach is called 'top-down,' because it is based on atmospheric measurements and an inversion model. Prather et al (2012) provide a spreadsheet model; here we employ the one-box mixing model of Daniel et al (2007):

$$E = \frac{m_{i+1} - m_i e^{-t/r}}{f\tau(1 - e^{-t/r})} \tag{1}$$

where E is annual emissions (Tg N_2O-N yr^{-1}), m_{i+1} and m_i are the observed atmospheric mixing ratios (ppb N_2O) at the start of consecutive years, τ is the lifetime, t is 1 year, and f relates the mass burden to the mixing ratio (0.21 ppb N_2O/Tg N; Prather et al 2012).

Estimates of atmospheric N_2O mixing ratios (specifically, the tropospheric mean mole fraction) prior to the industrial revolution are from ice cores measurements, which were at an approximate steady state from 1730 to 1850 (Machida et al 1995). An important source of uncertainty in the top-down approach is the estimated atmospheric lifetime of N_2O. Most estimates range between the IPCC-AR4 assumption of 114 years (Forster et al 2007) to 131 (Prather et al 2012). The Stratosphere-troposphere Processes And their Role in Climate report (SPARC 2013) suggests 123 years as the

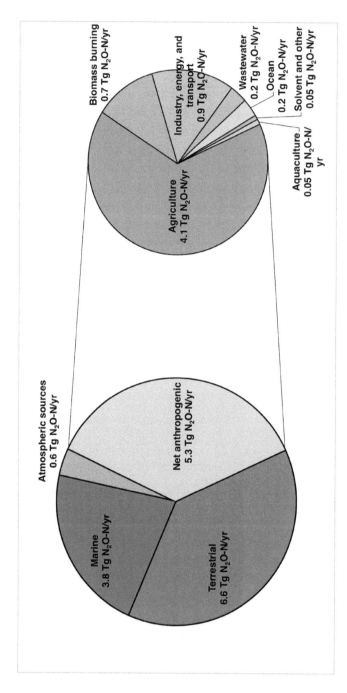

Figure 1. Natural versus anthropogenic N_2O emissions in 2005. The values for natural emissions (terrestrial, marine, and atmospheric chemistry in the pie chart on the left) are taken from Ciais et al (2013), while the anthropogenic values are the best estimate values by sector from the 2013 UNEP report (Bouwman et al 2013, Oenema et al 2013, van der Werf et al 2013, Wiesen et al 2013), as summarized in table 1 of this paper. The net anthropogenic estimate in the left pie chart takes into account the effect of land use change on reducing net anthropogenic emissions (about 0.9 Tg N_2O-N yr^{-1}, see section 3.2). Reprinted, with permission, from Davidson et al (2013), (figure 3.1).

'recommended' estimate, with the 'most likely range' between 104 and 152 years. Because of this large range in estimates from a number of respected sources, we choose here to use only two significant figures. We round the SPARC recommended estimate to 120 years, with an uncertainty range 110–130 years, which encompasses most of the range of central estimates used commonly in the literature. We also assume that the lifetime has not changed substantially, due to a lack of compelling evidence to the contrary. With these assumptions, pre-industrial emissions are estimated to be about 11 Tg N_2O-N yr^{-1}, with an uncertainty range of 10–12 Tg N_2O-N yr^{-1}. The sensitivity of the estimate is a change of about 1 Tg N_2O-N yr^{-1} of pre-industrial emissions for every 10 years change in assumed lifetime. Hence, if the lifetime was as long as about 140 years in the pre-industrial period, the emissions estimate would be 9 Tg N_2O-N yr^{-1} (Prather et al 2012). The central estimates of both top-down and bottom up approaches for pre-industrial natural emissions are in agreement at 11 Tg N_2O-N yr^{-1}, although both have considerable uncertainties.

Uncertainty in pre-industrial natural emission estimates also arises from the lack of complete understanding about the influence of anthropogenic changes prior to the industrial revolution (such as the expansion of agriculture) and from temporal variability of natural emissions. Between 1730 and 1850, N_2O concentrations in the atmosphere varied slightly from year to year and decade to decade, but showed little or no consistent long term trend (Machida et al 1995). Although Syakila and Kroeze (2011) estimated that average net anthropogenic emissions were around 0.5 Tg N_2O-N yr^{-1} during the 18th and early 19th centuries, these possible anthropogenic emissions were sufficiently low that their signal cannot be distinguished from the effects of climatic variation on natural emissions.

Despite the uncertainties, both bottom-up and top-down approaches suggest that natural emissions were and probably still are between 10 and 12 Tg N_2O-N yr^{-1}. We will concentrate the rest of our analyses on anthropogenic effects since 1850.

7.3 ANTHROPOGENIC EMISSIONS

7.3.1 TOP-DOWN ATMOSPHERIC MODELING CONSTRAINTS

Modern anthropogenic emissions of N_2O can be calculated using the same top-down method described above (equation (1)). In this case, the changes in atmospheric concentrations from 1850 to the present (Machida et al 1995, NOAA 2014) are assumed to be entirely anthropogenic, assuming relative stability of natural emissions over the same period and the same atmospheric N_2O lifetime of 120 (±10) years. The natural emission estimate (11 Tg N yr^{-1}) is subtracted from the total modern emissions calculated from equation (1) to yield modern net anthropogenic emissions. We averaged the emission estimates for the period 2000–2007 to avoid artifacts of short-term interannual variation, yielding an estimate for net anthropogenic emissions of 5.3 Tg N_2O-N yr^{-1} (range 5.2–5.5) for that period. This estimate includes all anthropogenic activities that have contributed to changing atmospheric N_2O, including any decrease in emissions from forest soils because of deforestation and increases in emissions from expanded activity in agriculture and other sectors.

7.3.2 BOTTOM-UP INVENTORY ESTIMATES

Protocols have been developed by the IPCC (2006) for countries to estimate their N_2O emissions. The IPCC Tier 1 Protocol multiplies metrics of activity in agriculture, energy generation, transportation, and other sectors, by emission factors (EFs), the amount of N_2O emitted per unit of activity. For example, the direct emissions of N_2O from agricultural soils are estimated as a 1% EF applied to synthetic-N fertilization application activity rates. Additional EFs are used to calculate the amount of fertilizer N leached into surface and groundwaters and volatilized as ammonia or nitrogen oxide gases, and the subsequent indirect N_2O emissions from downstream and downwind ecosystems, which often are substantial. For example, emissions from coastal, estuarine and riverine waters are estimated to be about 9% of total anthropogenic sources (Ciais et al 2013), although the original source of most of this N was from agricultural field applications. The EFs have been derived from the literature and are periodically revised as warranted. By necessity, they are averages across a broad range of conditions and often do not yield accurate estimates for individual sites. Nevertheless, there is evidence that errors on the small scale are largely canceled when aggregated to larger scales (Del Grosso et al 2008).

Another source of inaccuracy in the use of Tier 1 EFs is that they assume a linear relationship between N application rates and N_2O emissions. A growing number of studies demonstrate nonlinear (usually exponential) relationships between N application rate and N_2O emissions (Shcherbak et al 2014). The nonlinear relationship is likely the result of large increases in N_2O emissions once N application rates are in excess of plant demands. This has important implications for targeting mitigation where N application rates are higher than N harvested in crop export and for not discouraging additional N application in N-deficient regions where mining of soil nutrients is common, such as sub-Saharan Africa. The implications of nonlinearity are not yet clear for global N_2O budgets. The differences between linear and nonlinear models for estimating N_2O emissions are more likely to be important at the farm scale compared to the global scale, because the biases of the linear model (probably overestimation of fluxes where N applications are low and underestimation where N application rates are high) at least partially cancel as the spatial scale increases.

The United Nations Food and Agriculture Organization (FAO) estimates agricultural N_2O emissions by applying IPCC Tier 1 EFs to their country data gathered from national publications and questionnaires. The Emissions Database for Global Atmospheric Research (EDGAR) database uses a blend of private and public data, applying IPCC Tier 1 EFs to estimate both non-agricultural and agricultural N_2O emissions (with the exception of biomass burning, where they apply EFs described in Andreae and Merlet (2001).

A variant of the bottom-up global inventory approach involves a combination of 'top-down' constraints, based on a global atmospheric budget, and 'bottom-up' inventory estimates of minor N_2O sources from biomass burning, industry, energy, and transportation sectors, and from statistical correlations at the global scale using data on fertilizer use, manure production, and land-use change (Crutzen et al 2008, Davidson 2009, Smith et al 2012). These approaches yield EFs based on newly fixed N (either Haber–Bosch or biological N fixation) and N remobilized from tillage of soils (Smith et al 2012) or through production of manure by livestock (Davidson 2009). They implicitly include both direct and indirect emissions (i.e., on the farm and downwind and downstream) from these N fluxes, so comparison with the IPCC EFs is not straightforward. Nevertheless, the estimates from Davidson (2009) of 2.0% of manure-N and 2.5% of fertilizer-N converted to N_2O are not far off of the sums of IPCC EFs for direct and indirect agricultural emissions and human sewage. The estimate from Smith et al (2012) that 4% of newly fixed N is converted to N_2O may not be far off of the sum of IPCC EFs when the cascading effects of newly fixed N moving through croplands, livestock operations, downwind and downstream ecosystems, and human sewage are considered.

Countries that have sufficient data to calculate EFs more specific to their particular situations are allowed to use them under IPCC's Tier 2 Protocol, which presumably yields more accurate estimates for those specific regions and management practices (IPCC 2006). Under the Tier 3 Protocol, countries with access to validated biogeochemical models and sufficient input data are allowed to use these models to calculate N_2O emissions (IPCC 2006). This presumably yields even more accurate estimates if the models skillfully account for spatial and temporal variation of the most important factors affecting emissions.

With the advent of new laser technologies for measurements of N_2O fluxes (e.g., Savage et al 2014) there is likely to be continued improvement in estimating emission factors for the Tier 1 and Tier 2 Protocols and for developing and validating the biogeochemical models used with the Tier 3 Protocol. However, it will remain difficult to fully account for the large spatial and temporal variation of N_2O emissions. Improvement in the quality of activity data for each country, such as its fertilizer application rates, livestock production, and manure handling procedures, is also necessary for improved emission estimates. New EFs are also needed for new cropping systems, such as second generation biofuel crops. Indeed, fertilizer application rates and EFs for biofuel production are among the largest uncertainties for projections of future N_2O emissions (see section 5.2).

Table 1 summarizes recent efforts at partitioning anthropogenic emissions from bottom-up inventories and from integrated bottom-up and top-down analyses. Here we adopt the recent estimates from UNEP (2013) for total net anthropogenic N_2O emissions of 5.3 Tg N_2O-N yr^{-1}, which is equal to the top-down estimate (section 3.1). The 'best estimate' from the UNEP report is lower than the estimates from other inventories shown in table 1, partly because of some lower updated sectoral estimates and partly due to including the effect of lower tropical forest soil emissions resulting from historic and on-going deforestation, which is neglected in many other inventories. The best estimate of gross anthropogenic emissions is 6.2 Tg N_2O-N yr^{-1}. Because tropical forest soils are a large natural source of N_2O emissions, tropical deforestation should be considered as a significant human-induced decrease in emissions. Soil N_2O emissions from recently converted tropical forests may initially increase, but the long-term trend is for emissions from the pasture soils and degraded land soils to be lower than those from intact, mature tropical forests (Davidson et al 2001, Melillo et al 2001), resulting in current estimates of a decreased source of 0.9 Tg N_2O-N yr^{-1} (Davidson 2009). Subtracting the effect of tropical deforestation from the estimate of gross anthropogenic emissions yields a best estimate of 5.3 Tg N_2O-N yr^{-1} for net anthropogenic emissions, which is 15% below the gross anthropogenic emission estimate. Without this adjustment, the bottom-up and top-down approaches would not agree, although the apparent exact agreement to a tenth of a teragram is probably partly fortuitous.

7.3.3 ANTHROPOGENIC EMISSIONS BY SECTOR

Significant uncertainties remain for activity data and especially for several of the emission factors in each sector. Brief summaries of expert analyses from each sector from chapters 4–7 of the UNEP (2013) report are presented here.

7.3.3.1 AGRICULTURE

Agriculture is the largest source of anthropogenic N_2O emissions, responsible for 4.1 Tg N_2O-N yr^{-1} (3.8–6.8 Tg N_2O-N yr^{-1}; Oenema et al 2013) or 66% of total gross anthropogenic emissions (table 1). Emission estimates include direct soil emissions from synthetic N fertilizer and manure application and indirect emissions that occur from downstream or downwind water bodies and soils after nitrate leaches away from croplands and after N emitted from croplands as ammonia or nitrogen oxide gases fall back to earth as atmospheric N deposition. Also included are N_2O emissions resulting from crop residues, manure management, cultivation of organic soils, and crop biological N fixation (C-BNF). The central factor responsible for agricultural N_2O emissions is a lack of synchronization between crop N demand and soil N supply, with, on average, around 50% of N applied to soils not being taken up by the crop (Snyder et al 2009, Oenema et al 2013, Venterea et al 2012). Inputs of N to agricultural soils are mainly from synthetic N fertilizer and manure application, with additional supply from legume N fixation, crop residues, and N deposition.

7.3.3.2 INDUSTRY AND FOSSIL FUEL COMBUSTION

The industry sector plus fossil fuel combustion (stationary combustion and transportation) together are responsible for about 0.9 Tg N_2O-N yr^{-1} (0.7–1.6 Tg N_2O-N yr^{-1}) or 15% of total gross anthropogenic N_2O emissions (Wiesen et al 2013). Nitric and adipic acid production are the major industrial sources. Nitric acid is mainly used as a feedstock in the production of explosives and N fertilizer, particularly ammonium nitrate, with N_2O emitted during the ammonia oxidation process (Lee et al 2011). Adipic acid is a key feedstock in synthetic fiber production, with N_2O resulting from the use of nitric acid to oxidize several organic chemicals

Table 1. Published inventories of N_2O emissions by sector.

All units in Tg N_2O-N yr^{-1}	FAO[a]	EDGAR[b]	EPA (2012)[c]	Syakila and Kroeze (2011)	Davidson (2009)[d]	Del Grosso et al (2008)[e]	Crutzen et al (2008)[f]	Denman (2007)[g]	Mosier et al (1998)[h]	UNEP report[i]	Estimate range
Agriculture	4.1	3.8	4.6	5.1	5	4.9	4.3–5.8	5.1	6.8	4.1[j]	3.8–6.8
Fertilizer	1.4	—								—	
Direct	1.1				2.2					—	
Indirect	0.3	3.6		2.3					3.9	—	
Manure	1.8									—	
Direct	1.4		4.2		2.8	4.5				—	
Indirect	0.4									—	
Organic soils	0.2	—	0.1	0.1	—	—	—	—	0.1	—	
Crop residues	0.3	—	0.3	0.3	—	—	—	—	0.4	—	
C-BNF	—	—	0.1	0.1	—	—	—	—	0.1	—	
Manure management	0.3	0.2	0.4	2.3	—	0.4	—	—	2.3	—	
Biomass burning	—	1.1[k]	1.7	0.7	0.5	—	—	0.7	—	0.7	0.5–1.7
Residue burning	0.01	—	1.6	—	—	—	—	—	—	0.04	
Savanna burning	—	—	—	—	—	—	—	—	—	0.3	
Other[i]	—	—	0.1	—	—	—	—	—	—	0.4	

Table 1. Continued.

Industry, energy and transportm	—	1.7	0.9	—	0.8	—	0.7–1.3	0.7	1.3	0.9	0.7–1.7
Wastewater	—	0.2	0.2	0.3	—	—	—	0.7	0.3	0.2	—
Aquaculture	—	—	—	—	—	—	—	—	—	0.05	—
Solvent and other product use	—	—	0.2	—	—	—	—	—	—	0.05	—
Land-use change[e]	—	—	—	—0.6	—0.9	—	0–0.9	—	—	—0.9	—
Ocean	—	—	—	1.0	—	—	—	—	—	0.2[o]	—
Total	—	6.8	7.6	6.5	5.4	—	5.6–6.5	6.7	8.4	5.3	5.3–8.4

[a] 2010 estimates; UN Food and Agriculture Organization (FAO) estimates agricultural N_2O emissions by applying IPCC Tier 1 EFs to their country data gathered from national publications and questionnaires. [b]2008 estimates; the Emissions Database for Global Atmospheric Research (EDGAR) category 'Indirect N_2O from non-agricultural NH_3 and NOx' is partitioned between 'Biomass Burning' and' Industry, Energy, Transport' and weighted by their total direct emissions. The EDGAR database uses a blend of private and public data, applying IPCC Tier 1 EFs to estimate both non-agricultural and agricultural N_2O emissions (with the exception of biomass burning, where they apply EFs described in Andreae and Merlet 2001). [c]2006 estimates. [d]2005 estimates; agriculture estimates include human waste emissions. [e]Mix of data from 2000 and 1994. [f]2000 estimates; agriculture estimates include human waste, aquaculture, and biomass burning emissions. [g]Effects of atmospheric deposition are included in the agriculture sector, although a portion of the deposition comes from other sectors. [h]2006 estimates for agriculture and waste (adapted by Syakila and Kroeze 2011). [i]1989 estimates for industrial/energy. [j]The UNEP estimates for agriculture, biomass burning, wastewater, aquaculture,

Table 1. Continued.

and land use change are from Bouwman et al (2013), Oenema et al (2013), van der Werf et al (2013), Wiesen et al (2013). ʲThe FAO estimates are adopted by UNEP (Oenema et al 2013). This estimate does include indirect emissions from downwind and downstream ecosystems, but does not include sewage wastewater emissions. ᵏIncludes category 'large-scale biomass burning' which denotes savanna burning, forest fires, peat fires, grassland fires, decay of wetland/peatland and post burn decay after forest fires, agricultural residue burning, and other vegetation fires. ˡ'Other' biomass burning includes tropical, temperate and boreal forest fires, tropical peat fires, and fuelwood fires. ᵐSeveral literature sources combine emissions from industry and/or energy and transport into one overall estimate. ⁿThe S&K estimate of reduced natural emissions is for pre-industrial land use change only. Crutzen et al provide only a range, so we use Davidson's (2009) estimate for post-industrial tropical deforestation. ᵒ Emissions from the ocean due to anthropogenic N deposition should be included in indirect emission factors for agriculture and other sectors, but are probably underestimated, so we include this estimate from Suntharalingam et al (2012).

(Schneider et al 2010). Stationary combustion (mainly coal power plants) is the principal source of N_2O from the energy sector. Emissions of N_2O from this sector arise via the oxidation of both atmospheric N_2 and organic N in fossil fuels. Emissions vary with the amount of organic N in the fuel, the operating temperature, and the oxygen levels during combustion (EPA 2012). N_2O from transport is released primarily by catalytic converters used to control NOx, carbon monoxide, and hydrocarbons in tailpipe emissions, with older technologies responsible for significantly higher emission rates per kilometer than more advanced technologies (IPCC 2006).

7.3.3.3 BIOMASS BURNING

Biomass burning is currently responsible for about 0.7 Tg N_2O-N yr⁻¹ (0.5–1.7 Tg N_2O-N yr⁻¹; van der Werf et al 2013) or 11% of total gross anthropogenic emissions. This includes crop residue burning, forest fires (resulting from both natural and human activities), and prescribed savannah, pasture, and cropland burning. It also includes N_2O emissions from household biomass stoves. N_2O is released via the oxidation of organic N in biomass during combustion. Although some wildfires are ignited naturally by lightning, all emissions from biomass burning have been attributed as anthropogenic emissions, because it is impossible to separate out which wildfires are ignited by humans. Furthermore, anthropogenic climate change may also be increasing fire frequency and severity (Pechony and Shindell 2010).

7.3.3.4 WASTEWATER, AQUACULTURE, AND OTHER SOURCES

N_2O emissions from wastewater were 0.2 Tg N_2O-N yr^{-1} in 2010 (Bouwman et al 2013), or 3% of total gross anthropogenic emissions. This includes N_2O emitted either directly from wastewater effluent or from bioreactors removing N in biological nutrient removal plants (Law et al 2012). A small amount of N_2O is also emitted in aquaculture (<0.1 Tg N_2O-N yr^{-1} in 2010). Various human-related changes to the oceanic environment have affected the amount of N_2O emissions produced by the oceans. Increased N deposition onto the ocean has been estimated to have increased the oceanic N_2O source by 0.2 Tg N_2O-N yr^{-1} (0.08–0.34 Tg N_2O-N yr^{-1}) or 3% of total gross anthropogenic emissions (Suntharalingam et al 2012). In principle, increased oceanic emission due to N deposition should be included in the indirect emission estimates from agricultural, energy, and transportation sources, but it is included here as a separate category because the oceans may have been under-represented in calculations of emissions from downwind and downstream ecosystems.

7.4 TRENDS IN EMISSIONS OVER THE LAST 20 YEARS

Figure 2 illustrates how N_2O emissions from three of the most important sectors of the EDGAR (2009) database have changed from 1990 to 2008. The dominance of emissions by agricultural soils is clear, with the importance of South Asia, parts of Latin America and especially East Asia growing in the last two decades. Large-scale biomass burning emissions are most important in tropical savannah regions. Industrial emissions are most important in developed countries and are growing in South and East Asia.

7.5 EMISSION PROJECTIONS

7.5.1 SYNTHESIS OF PUBLISHED SCENARIOS

Projections of future emissions depend upon assumptions about changes in:

- Population growth rates.
- Per capita consumption of calories and protein.
- Relative sources of vegetable versus animal products for meeting food demands.
- Rates of wastage/loss of food from production to consumption.
- Nutrient use efficiency in crop and animal production systems.
- Production of newly fixed N for agriculture (including biofuels) and aquaculture.
- Emissions of NH_3 and NOx from all sectors, which contribute to N deposition on native soils and oceans.
- Fire frequency, including household biomass burning, slash-and-burn agriculture, pasture clearing, and wildfire.
- Industrial and energy sectors (such as fertilizer manufacturers and industries using coal combustion) that can reduce emissions.
- Land-use change.
- Energy sector technology and demand for biofuels.
- Climate and its effects on N cycling processes.

Climate change can affect N_2O emissions from water bodies and soils under native vegetation, but this effect is not well represented in current models (Pinder et al 2012) and it is not dealt with here. Most published projections of future emissions focus on assumptions about changes in emissions from agriculture, biomass burning, energy, transportation, and industry, which vary widely among the scenarios considered here (table 2) and elsewhere (e.g. Bodirsky et al 2014). Here, four sets of published N_2O emission scenarios were aggregated to characterize the potential range of future anthropogenic emissions:

- The Special Report on Emissions Scenarios (SRES) (Nakicenovic et al 2000) created four major global greenhouse gas emissions

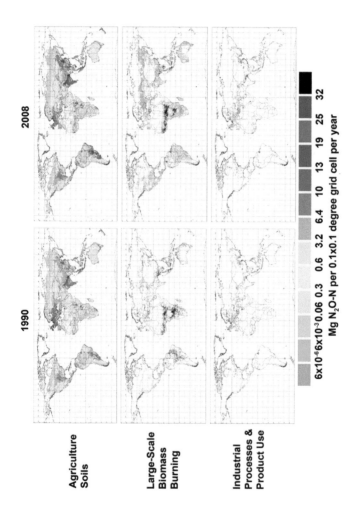

Figure 2. Global maps of direct emissions from agricultural soils, large scale biomass burning, and industrial process (mostly adipic and nitric acid production) for 1990 and 2008 from the Emissions Database for Global Atmospheric Research (EDGAR), version 4.0 (2009) (http://edgar.jrc.ec.europa.eu). Units are tons of N_2O-N per 0.1 º 0.1 degree grid cell per year. Note that this size grid cell is about 123 km² at the equator and declines with increasing latitude, approaching zero near the poles. Reprinted, with permission, from Davidson et al (2013), (figure 3.2).

scenarios (A1, A2, B1 and B2) based on the degree of globalization versus regionalization and the priority given to economic versus social and environmental objectives. These were used in the IPCC's Third and Fourth Assessment Reports.

- The RCPs (Van Vuuren et al 2011a) are used in the IPCC AR5, with four scenarios based on differing radiative forcing levels rather than emissions (RCP 2.6, 4.5, 6.0 and 8.5, with the numbers referring to different radiative forcing levels in Wm−2 in the year 2100).

- Davidson (2012) used FAO projections of population and dietary demands to estimate fertilizer and manure demands and subsequent N_2O emissions, including five variants (S1-S5) of mitigation and dietary habits.

- Five scenarios of a new UNEP report (Sutton et al 2013) based on expert analyses of feasible mitigation options in each sector: TR1: Business-As-Usual; TR2: Mitigation of Industry, Fossil Fuel Combustion, and Biomass Burning; TR3: Efficiency of Agricultural Production; TR4: Efficiency of Agricultural Production and Consumption; and TR5: Combined Mitigations.

These studies have different base years and employ different inventory sources. In order to make their results comparable, all emission estimates were normalized to the best estimate of 2005–2010 average net anthropogenic emissions from the UNEP report (5.3 Tg N_2O-N yr^{-1}). The scenarios of annual emissions are presented graphically in figure 3. The numerous scenarios are organized into three groups and means calculated for each group:

7.5.2 BUSINESS-AS-USUAL SCENARIOS (BAU)

The RCP 8.5, SRES A2, Davidson's S1, and UNEP1 scenarios have no or little mitigation. On average, the emissions of these scenarios increase to 9.7 Tg N_2O-N yr^{-1} by 2050, which is nearly double their level in 2005 (83% increase).

Table 2. Summary of assumptions of published scenarios for future N_2O emissions.

Source	Scenario	Description
SRES (Nakicenovic et al 2000)	A1	A world of increased regional, economic, social and cultural convergence drives rapid economic growth and the dissemination of new technologies, with global population peaking at nine billion in 2050 and declining thereafter.
	A2	A fragmented world with more regionally focused economic development leads to slower per capita economic growth and technological change than other scenarios. Continuous population growth due to slow convergence of regional fertility patterns.
	B1	A global approach to economic, social and environmental sustainability leads to the swift creation of a service and information economy, with a rapid expansion of clean technologies and less resource use. Similar population growth to A1.
	B2	A similar commitment to sustainability as in B1, but with an emphasis on local solutions. Continuous population growth (at a slower rate than A2), moderate levels of economic development, with slower and more varied technological change than A1 and B1.
RCPs (van Vuuren et al 2011a)	RCP8.5	A high emissions pathway—representing a scenario where little is done to limit climate change—leading to radiative forcing of 8.5 Wm^{-2} in 2100.
	RCP6	An emissions pathway that eventually leads to a stabilization of radiative forcing at 6 Wm^{-2} after 2100.
	RCP4.5	An emissions pathway that eventually leads to a stabilization of radiative forcing at 4.5 Wm^{-2} after 2100.
	RCP2.6	A low emissions pathway—representing ambitious international action to limit climate change—that leads to a peak in radiative forcing at 3 Wm^{-2} before 2100, dropping to 2.6 Wm^{-2} by 2100.
Davidson (2012)	S1	Future fertilizer and manure use scaled to FAO projections of population growth, per capita caloric intake and meat consumption.

Table 2. Continued.

	S2	Developed countries reduce per capita meat consumption to 50% of 1980 levels by 2030, remaining constant to 2050. Results in a 21% reduction in fertilizer use and manure production by 2030 and 2050 relative to Scenario 1.
	S3	Improved efficiency of fertilizer use and manure production reduces N2O emission factors 50% by 2050.
	S4	The same as Scenario 3, with the addition of emission reductions from industry, energy, transport, and biomass burning of 50% by 2050.
	S5	A scenario combining the mitigation actions of Scenarios 2 through 4.
UNEP (Sutton et al 2013)	TR1	Emissions are projected to increase according to the 'business-as-usual' assumptions presented in Chapters 4 to 7 of UNEP (2013).
	TR2	Combined emissions from industry, energy, transport and biomass burning are reduced by 58% by 2050 relative to Case 1.
	TR3	Improved efficiency of fertilizer use reduces fertilizer demand by 15% and the N_2O emission factor for fertilizer 20% by 2050 relative to Case 1. Improvements in manure management reduce N excretion per unit animal product by 30% and the N_2O emission factor for manure production 10% by 2050 relative to Case 1.
	TR4	The efficiency improvements in Case 3 are combined with 50% reductions in global food waste and developed country meat consumption relative to Case 1.
	TR5	A combination of all the mitigation actions in Cases 2 through 4.

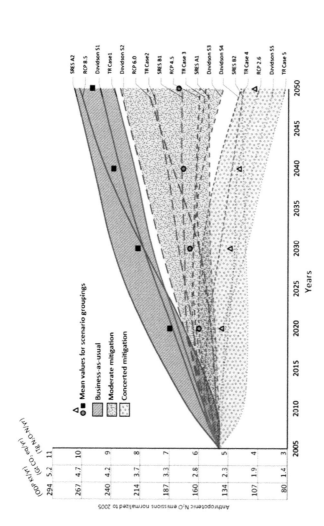

Figure 3. Projections of anthropogenic N$_2$O emissions according to groupings of published business-as-usual, moderate mitigation, and concerted mitigation scenarios (Nakicenovic et al 2000, Van Vuuren et al 2011a, Davidson 2012, Sutton et al 2013; see table 2). The 'TR' cases 1–5 refer to 'this report', being the 2013 UNEP report (Sutton et al 2013). The mean for each grouping of scenarios is shown by square, circle, and triangle markers. All projections have been adjusted to a common emission estimate baseline in 2005 consistent with the UNEP report's best estimate of net anthropogenic emissions of 5.3 Tg N$_2$O-N yr^{-1}. Concerted scenarios include aggressive mitigation in all sectors and most include reduced per capita meat consumption in the developed world. Y-axis units have been converted to CO$_2$ equivalents for a 100-year global warming potential and to Ozone Depletion Potential (ODP; Ravishankara et al 2009). Reprinted, with permission, from Sutton et al (2013), (figure 8.1).

7.5.3 MODERATE MITIGATION SCENARIOS

The scenarios RCP 4.5, RCP 6.0, SRES A1, SRES B1, Davidson's S2 and S3, UNEP2 and UNEP3 have moderate mitigation, defined here as scenarios showing emission trends that are higher than 2005 emissions in 2050 but below BAU. On average, emissions grow to 6.7 Tg N_2O-N yr^{-1} by 2050, an increase of 26% relative to 2005.

7.5.4 CONCERTED MITIGATION SCENARIOS

The RCP 2.6, SRES B2, Davidson's S4 and S5, UNEP4, and UNEP5 mitigation scenarios are concerted, because they lead to emissions in 2050 that are below the 2005 level. On average, emissions decline to 4.2 Tg N_2O-N yr^{-1} by 2050, a decrease of 22% relative to 2005.

The concerted mitigation scenarios result in near stabilization of atmospheric concentrations of N_2O between 340 and 350 ppb by 2050 (Davidson 2012, Davidson et al 2013), whereas N_2O concentration continues rising beyond 2050 for the BAU and moderate mitigation scenarios.

By 2020 the average concerted mitigation scenario reduces emissions by 1.8 Tg N_2O-N yr^{-1} or 25% below the average BAU scenario, equivalent to 0.8 Gt CO_2 eq yr^{-1} less than BAU (table 3). This is approximately 10% of the emissions gap that needs to be bridged by 2020 for it to be 'likely' that average global warming stays below a 2 °C threshold (UNEP 2012). By 2050, the average concerted mitigation scenarios are 57% lower than the average BAU scenario—around 5.5 Tg N_2O-N yr^{-1} (2.6 Gt CO_2 eq yr^{-1}). The avoided emissions between 2014 and 2050 sum to 22 Tg N_2O-N (57 Gt CO_2eq). To put this figure in context, it is equal to about ten years of the CO_2 emissions of all of the passenger cars currently on the road.

The impact of N_2O on stratospheric ozone depletion has been estimated using an ozone-depletion potential (Ravishankara et al 2009; ODP—a measure of its ozone destructiveness relative to CFC-11, which is defined as an ODP of 1). Although the use of ODPs is controversial because of complex interactions of various anthropogenic gases and stratospheric temperature (Fleming et al 2011), we employ it here to place the potential

Table 3. Projected annual anthropogenic N_2O emissions for three emission scenario groupings, given in units of N, CO_2 equivalents, and ozone depletion potential.

	2020	2030	2040	2050
Units: nitrogen equivalents (Tg N_2O-N yr^{-1})[a]				
Business-as-usual	7.0	8.1	8.9	9.7
Moderate mitigation	6.0	6.3	6.5	6.7
Concerted mitigation	5.2	5.0	4.7	4.2
Units: equivalents of carbon dioxide (Gt CO_2-eq yr^{-1})[b]				
Business-as-usual	3.3	3.8	4.2	4.5
Moderate mitigation	2.8	3.0	3.1	3.1
Concerted mitigation	2.5	2.3	2.2	1.9
Units: ozone depletion potential (ODP kt yr^{-1})[c]				
Business-as-usual	187	216	238	258
Moderate mitigation	160	169	175	178
Concerted mitigation	140	133	125	111

[a] The values are the mean of four sets of scenarios according to SRES (Nakicenovic et al 2000), RCP (van Vuuren et al 2011a), Davidson (2012), and UNEP (2013) and grouped as described in the main text. [b] Calculated using a 100-year global warming potential of 298 for N2O. [c] Calculated using an ozone depleting potential of 0.017 for N_2O.

impact of N_2O mitigation on stratospheric ozone in context with previous efforts to mitigate CFCs. By 2050, the difference between the concerted average mitigation scenario and the average BAU scenario (147 ODP kt yr^{-1} is equivalent to a 13% decrease in chlorofluorocarbon (CFC) emissions from their peak in the late 1980s, approximately halving ODP-weighted emissions in 2050 compared to BAU (table 3). The sum of the avoided emissions between 2014 and 2050 is 3270 ODP kt. These reductions would be 40%–110% greater than the potential reductions from the destruction of the remaining recoverable stocks of other ozone depleting substances, which has been identified as the most substantive remaining action that could be taken to accelerate ozone layer recovery (UNEP 2010).

An important caveat of these projections is that they all begin in 2005, and significant differences are already apparent in their trajectories by 2013. So far (up to 2013), actual global N_2O emissions have been closer to BAU trajectories than the mitigation trajectories.

7.5.5 PROJECTING N_2O EMISSIONS FROM BIOFUEL PRODUCTION

Another caveat of these projections is that the highly uncertain impact of expansion of biofuel production is not considered (Davidson et al 2013). In addition to uncertainties about total biofuels produced, the N fertilization rates needed for producing second- or third-generation fuel stocks and the N_2O EFs for those cultivation practices are not known. Fertilization rates and EFs for rapidly growing trees and native grasses, forbs, and shrubs may be much lower than for most current food and fiber crops. To put this uncertainty into perspective, two methods are offered here to bound the range of future N_2O emissions from biofuels—one based on the potential for energy production and the other based on total land available for biofuel crops.

For the first method, Edenhofer et al (2011) estimate a bioenergy deployment range of 100–300 EJ yr^{-1} by 2050, which takes into account soil conservation and biodiversity goals, as well as potential water scarcity and the use of land for subsistence farming (Edenhofer et al 2011, Creutzig et al 2012). For this calculation, it was assumed that by 2050 all bioenergy demand will be supplied by second-generation biofuels. Given data constraints, the estimation focuses on jatropha (*Jatropha curcas*), miscanthus (*Miscanthus x giganteus*), eucalyptus (*Eucalyptus cinera*) and switchgrass (*Panicum virgatum L*). To estimate emissions, a range of published N_2O EFs for these biofuels (0.2 to 27.1 g N_2O-N kJ^{-1}; Hoefnagels et al 2010) was used. This approach generates estimates of 0.02-8.1 Tg N_2O-N yr^{-1} from biofuels by 2050, depending on the fuel source and the total amount of bioenergy deployed, with a central estimate of 2.1 Tg N_2O-N yr^{-1} based on the combined means of the bioenergy deployment range and the published emission factors for second-generation biofuels.

The second method focuses on the amount of land potentially available to cultivate biofuel crops. Estimates range from 60–3700 Mha, covering 0.4%–28% of the Earth's land surface, excluding Greenland and Antarctica, with several estimates clustering between 240–500 Mha (Creutzig et al 2012). In comparison, Melillo et al (2009) estimated that 2000 Mha of biofuel crop cultivation will be needed by 2100 to stabilize atmospheric CO_2 concentrations at 550 ppm, while van Vuuren et al (2011b) estimated that 3000–4000 Mha will be needed by 2100 in the RCP 2.6 scenario. We assume an average fertilizer application rate of 100 kg N h^{-1} yr^{-1} for land devoted to biofuel crops, as was done by Erisman et al (2008), and use the IPCC (2006) direct and indirect default EFs. Using this approach, N_2O emissions were estimated to be 0.08–4.9 Tg N_2O-N yr^{-1}, depending on the amount of land devoted to biofuel crop cultivation, with a central estimate of 0.5 Tg N_2O-N yr^{-1} based on the mean of the cluster of land-use estimates cited in Creutzig et al (2012).

These estimates are considerably lower than the 16.1–18.6 Tg N_2O-N yr^{-1} estimated by Melillo et al (2009) for 2100. These data illustrate the huge uncertainty that still remains in future estimates of N_2O emissions from biofuels. Comparing these estimates from <1 to 18 Tg N_2O-N yr^{-1} to the range of the aggregated RCP, SRES and Davidson (2012) scenarios (4.4 to 9.9 Tg N_2O-N yr^{-1}, table 3) demonstrates that biofuels could either remain a relatively trivial source or become the most significant source of anthropogenic N_2O emissions at some point this century. Energy and climate policy decisions in the coming decades as well as the pace of technical innovation will be among the major determinants of future N_2O emissions from biofuels.

7.6 CONCLUSIONS

- Natural N_2O emissions are most likely between 10 and 12 Tg N_2O-N yr^{-1}.
- Both bottom-up and top-down analyses suggest that net anthropogenic N_2O emissions are now (2005–2010) about 5.3 Tg N_2O-N yr^{-1}.
- Agriculture currently accounts for 56–81% of gross anthropogenic N_2O emissions. Some N_2O emissions associated with food production is inevitable, but future N_2O emissions from agriculture will be determined by several factors, including population, dietary habits, and agricultural management to improve N use efficiency.
- The BAU emission scenarios project almost a doubling of anthropogenic N_2O emissions, from 5.3 Tg N_2O-N yr^{-1} in 2005 to 9.7 Tg N_2O-N yr^{-1} in 2050. In contrast, the concerted mitigation scenarios result in an average decline to 4.2 Tg N_2O-N yr^{-1} by 2050, a decrease of 22% relative to 2005, which would lead to a near stabilization of atmospheric concentration of N_2O at about 350 ppb.
- The impact of growing demand for biofuels on future N_2O emissions is highly uncertain, depending on the types of plants grown, their nutrient management, the amount of land dedicated to their cultivation, and the fates of their waste products. N_2O emissions from second and third generation biofuels could remain trivial or could become the most significant source to date. Research is needed to reduce the uncertainty of the future impact of biofuels on N_2O.

REFERENCES

1. Andreae M O and Merlet P 2001 Emission of trace gases and aerosols from biomass burning Glob. Biogeochem. Cycles 15 955–66
2. Bodirsky B L et al 2014 Reactive nitrogen requirements to feed the world in 2050 and potential to mitigate nitrogen pollution Nat. Commun. 5 3858
3. Bouwman L, van der Sluis S, Zhang G-L, Towprayoon S and Mulsow S 2013 Reducing N2O emissions from wastewater and aquaculture Drawing Down N2O

to Protect Climate and the Ozone Layer. A UNEP Synthesis Report (Nairobi: United Nations Environment Programme) (www.unep.org/pdf/UNEPN2Oreport. pdf)

4. Ciais P et al 2013 Carbon and other biogeochemical cycles Climate Change 2013: The Physical Science Basis. Contribution of Working Group I to the Fifth Assessment Report of the Intergovernmental Panel on Climate Change ed T F Stocker et al (New York: Cambridge University Press)

5. Creutzig F, Popp A, Plevin R, Luderer G, Minx J and Edenhofer O 2012 Reconciling top-down and bottom-up modelling on future bioenergy deployment Nat. Clim. Change 2 320–7

6. Crutzen P J, Mosier A R, Smith K A and Winiwarter W 2008 N2O release from agro-biofuel production negates global warming reduction by replacing fossil fuels Atmos. Chem. Phys. 8 389–95

7. Daniel J S, Velders G J M, Solomon S, McFarland M and Montzka S A 2007 Present and future sources and emissions of halocarbons: toward new constraints J. Geophys. Res. 112 D02301

8. Davidson E A 2009 The contribution of manure and fertilizer nitrogen to atmospheric nitrous oxide since 1860 Nature Geosci. 2 659–62

9. Davidson E A 2012 Representative concentration pathways and mitigation scenarios for nitrous oxide Environ. Res. Lett. 7 024005

10. Davidson E A, Bustamante M M C and Pinto A D S 2001 Emissions of nitrous oxide and nitric oxide from soils of native and exotic ecosystems of the Amazon and Cerrado regions of Brazil Optimizing nitrogen management in food and energy production and environmental protection: Proc. of the 2nd Int. Nitrogen Conf. on Science and Policy ed J Galloway et al (Lisse: A A Balkema Publishers)

11. Davidson E A, Kanter D, Suddick E and Suntharalingam P 2013 N2O: sources, inventories, projections Drawing Down N (Nairobi: United Nations Environment Programme) (www.unep.org/pdf/UNEPN2Oreport.pdf)

12. Del Grosso S J, Wirth T, Ogle S M and Parton W J 2008 Estimating agricultural nitrous oxide emissions EOS Trans. Am. Geophys. Union 89 529–40

13. Denman et al 2007 Couplings between changes in the climate system and biogeochemistry Climate change 2007: The Physical Science Basis. Contribution of working Group I to the Fourth Assessment Report of the Intergovernmental Panel on Climate Change (Cambridge: Cambridge University Press) ed S Solomon et al

14. Edenhofer O, Pichs-Madruga R, Sokona Y and Seyboth K 2011 IPCC Special Report on Renewable Energy Sources and Climate Change Mitigation (New York: Cambridge University Press)

15. EPA 2012 Global Anthropogenic Non-CO2 Greenhouse Gas Emissions (Washington DC: Environmental Protection Agency) 430-R-12-006

16. EDGAR 2009 Emissions database for global atmospheric research (EDGAR) version 4.0 (http://edgar.jrc.ec.europa.eu)

17. Erisman J W, van Grinsven H, Leip A, Mosier A and Bleeker A 2010 Nitrogen and biofuels; an overview of the current state of knowledge Nutr. Cycling Agroecosyst. 86 211–23

18. Firestone M K and Davidson E A 1989 Microbiological basis of NO and N2O production and consumption in soil Exchange of Trace Gases between Terrestrial Ecosystems and the Atmosphere ed M O Andreae and D S Schimel (New York: Wiley)

19. Fleming E L, Jackman C H, Stolarski R S and Douglass A R 2011 A model study of the impact of source gas changes on the stratosphere for 1850–2100 Atmos. Chem. Phys. 11 8515–41

20. Forster P et al 2007 Changes in atmospheric constituents and in radiative forcing Climate Change 2007: The Physical Science Basis. Contribution of Working Group I to the Fourth Assessment Report of the Intergovernmental Panel on Climate Change ed S Solomon et al (Cambridge: Cambridge University Press)

21. Galloway J N et al 2004 Nitrogen cycles: past, present and future Biogeochemistry 70 153–226

22. Galloway J N, Aber J D, Erisman J W, Seitzinger S P, Howarth R W, Cowling E B and Cosby B J 2003 The nitrogen cascade Bioscience 53 341–56

23. Hoefnagels R, Smeets E and Faaij A 2010 Greenhouse gas footprints of different biofuel production systems Renew. Sustainable Energy Rev. 14 1661–94

24. IPCC 2006 IPCC guidelines for national greenhouse gas inventories (www.ipcc-nggip.iges.or.jp/public/2006gl/)

25. Law Y, Ye L, Pan Y and Yuan Z 2012 Nitrous oxide emissions from wastewater treatment processes Phil. Trans. R. Soc. B 367 1265–77

26. Lee S-J, Ryu I-S, Kim B-M and Moon S-H 2011 A review of the current application of N2O emission reduction in CDM projects Int. J. Greenh. Gas. Con. 5 167–76

27. Machida T, Nakazawa T, Fujii Y, Aoki S and Watanabe O 1995 Increase in the atmospheric nitrous-oxide concentration during the last 250 years Geophys. Res. Lett. 22 2921–4

28. Melillo J M et al 2009 Indirect emissions from biofuels: how important? Science 326 1397–9

29. Melillo J M, Steudler P A, Feigl B J, Neill C, Garcia D, Piccolo M C, Cerri C C and Tian H 2001 Nitrous oxide emissions from forests and pastures of various ages in the Brazilian Amazon J. Geophys. Res. 106 179–88

30. Mosier A, Kroeze C, Nevison C, Oenema O, Seitzinger S and Van Cleemput O 1998 Closing the global N2O budget: nitrous oxide emissions through the agricultural nitrogen cycle Nutr. Cycling Agroecosyst. 52 225–48

31. Nakicenovic N et al 2000 Special Report on Emissions Scenarios: A Special Report of Working Group III of the Intergovernmental Panel on Climate Change (New York: Cambridge University Press)

32. NOAA 2014 Combined nitrous oxide data from the NOAA/ESRL Global Monitoring Division (ftp://ftp.cmdl.noaa.gov/hats/n2o/combined/HATS_global_N2O.txt)

33. Oenema O et al 2013 Reducing N2O emissions from agricultural sources A UNEP Synthesis Report (Nairobi: United Nations Environment Programme) Drawing Down N2O to Protect Climate and the Ozone Layer. (www.unep.org/pdf/UNEPN2Oreport.pdf)

34. Pechony O and Shindell D T 2010 Driving forces of global wildfires over the past millennium and the forthcoming century Proc. Natl. Acad. Sci. 107 19167–70

35. Pinder R, Davidson E A, Goodale C L, Greaver T L, Herrick J D and Liu L 2012 Climate change impacts of US reactive nitrogen Proc. Natl. Acad. Sci. 109 7671–5

36. Prather M J, Holmes C D and Hsu J 2012 Reactive greenhouse gas scenarios: systematic exploration of uncertainties and the role of atmospheric chemistry Geophys. Res. Lett. 39 L09803

37. Ravishankara A R, Daniel J S and Portmann R W 2009 Nitrous oxide (N2O): the dominant ozone-depleting substance emitted in the 21st century Science 326 123–5

38. Rockström J et al 2009 A safe operating space for humanity Nature 461 472–5

39. Savage K E, Phillips R and Davidson E A 2014 High temporal frequency measurements of greenhouse gas emissions from soils Biogeosciences 11 2709–20

40. Schneider L L, Lazarus M and Kollmus A 2010 Industrial N (Somerville, MA: Stockholm Environment Institute)

41. Shcherbak I, Millar N and Robertson G P 2014 A global meta-analysis of the nonlinear response of soil nitrous oxide (N2O) emissions to fertilizer nitrogen Proc. Natl. Acad. Sci. 111 9199–204

42. Smith K A, Mosier A R, Crutzen P J and Winiwarter W 2012 The role of N2O derived from biofuels, and from agriculture in general, in Earth's climate Phil. Trans. R. Soc. B 367 1169–74

43. Snyder C S, Bruulsema T W and Fixen P E 2009 Review of greenhouse gas emissions from crop production systems and fertilizer management effects Agric. Ecosyst. Environ. 133 247–66

44. SPARC 2013 SPARC Report on the Lifetimes of Stratospheric Ozone-Depleting Substances, Their Replacements, and Related Species ed M Ko et al SPARC report no. 6, WCRP-15/2013 (Geneva: World Climate Research Programme)

45. Suntharalingam P, Buitenhuis E, Le Quéré C, Dentener F, Nevison C, Butler J H, Bange H W and Forster G 2012 Quantifying the impact of anthropogenic nitrogen deposition on oceanic nitrous oxide J. Geophys. Res. 39 L07605

46. Sutton M A, Skiba U M, Davidson E A, Kanter D, van Grinsven H J M, Oenema O, Maas R and Pathak H 2013 Drawing-down N2O emissions: scenarios, policies and the green economy A UNEP Synthesis Report (Nairobi: United Nations Environment Programme) Drawing Down N2O to Protect Climate and the Ozone Layer. (www.unep.org/pdf/UNEPN2Oreport.pdf)

47. Syakila A and Kroeze C 2011 The global nitrous oxide budget revisited Greenhouse Gas Meas. Manage. 1 17–26

48. UNEP 2010 2010 Assessment Report of the Technology and Economic Assessment Panel (Nairobi: United Nations Environment Programme)

49. UNEP 2012 The Emissions Gap Report 2012. A UNEP Synthesis Report (Nairobi: United Nations Environment Programme)

50. UNEP 2013 A UNEP Synthesis Report (Nairobi: United Nations Environment Programme) Drawing Down N2O to Protect Climate and the Ozone Layer. (www.unep.org/pdf/UNEPN2Oreport.pdf)

51. van der Werf G R, Meyer C P and Artaxo P 2013 Reducing N2O emissions from biomass burning in landscape fires and household stoves A UNEP Synthesis

Report (Nairobi: United Nations Environment Programme) Drawing Down N2O to Protect Climate and the Ozone Layer. (www.unep.org/pdf/UNEPN2Oreport. pdf)

52. Van Vuuren D P et al 2011a The representative concentration pathways: an overview Clim. Change 109 5–31

53. Van Vuuren D P et al 2011b RCP2.6: exploring the possibility to keep global mean temperature increase below 2 C Clim. Change 109 95–116

54. Venterea R T et al 2012 Challenges and opportunities for mitigating nitrous oxide emissions from fertilized cropping systems Front. Ecol. Environ. 10 562–70

55. Wiesen P, Wallington T J and Winiwarter W 2013 Reducing N2O emissions from industry and fossil fuel combustion A UNEP Synthesis Report (Nairobi: United Nations Environment Programme) Drawing Down N2O to Protect Climate and the Ozone Layer.

PART IV

FOSSIL FUEL ALTERNATIVES

CHAPTER 8

Differences Between LCA for Analysis and LCA for Policy: A Case Study on the Consequences of Allocation Choices in Bio-Energy Policies

TJERK WARDENAAR, THEO VAN RUIJVEN,
ANGELICA MENDOZA BELTRAN, KATHRINE VAD,
JEROEN GUINÉE, AND REINOUT HEIJUNGS

8.1 INTRODUCTION

The increasing concern for possible adverse effects of climate change, has spurred the search for alternatives for conventional energy production systems. Biomass based energy (fuel, heat and electricity), or bio-energy, has in this respect been promoted as a promising alternative. Bio-energy is believed to be more sustainable than the conventional energies obtained from fossil fuels (Chum et al. 2011).

Moreover, it is believed that bio-energy increases countries' energy security and to create opportunities for rural development. As a consequence bio-energy is stimulated via environmental and energy policies in both

© The Author(s) 2012. The International Journal of Life Cycle Assessment, September 2012, Volume 17, Issue 8, pp 1059-1067. DOI 10.1007/s11367-012-0431-x. Creative Commons Attribution License (http://creativecommons.org/licenses/by/3.0/).

developed and developing countries (Worldwatch Institute 2007; United States Department of Energy 2010; Van der Voet et al. 2010).

Despite these advantages, bio-energy is increasingly linked to adverse effects on the environment and on society. Questions have been raised with respect to impacts on food, land and water availability (Bindraban and Pistorius 2008; De Fraiture et al. 2008). Another criticism concerns the alleged impacts on land use changes and the destruction of tropical rain forest (Searchinger et al. 2008). Also the presumed reductions in greenhouse gas (GHG) emissions are questioned (Reijnders and Huijbregts 2008). In a response to these more critical stances to bio-energy, governments have introduced directives with the intention to stimulate sustainable bio-energy (SenterNovem 2008; UNEP 2009). Life cycle assessment (LCA) plays an important role in these directives and it often serves as the main tool to assess alternative energy production systems' reductions in GHG emissions. In this way, policymakers are faced with methodological decisions central to LCA, e.g. with respect to the allocation method. This article reviews various bio-energy directives and discusses how their differences with respect to the recommended allocation methods may influence the assessment of bio-energy systems. It does so in order to stimulate the discussion on distinguishing LCAs for the purpose of analysis (finding hotspots, monitoring, process optimization, etc.) and LCAs for policy purposes (banning, subsidizing, certifying, etc.).

This paper is organized as follows. Section 2 sketches the issues of allocation and how it has been dealt with in policy guidelines on bio-energy. Section 3 describes a case study on electricity with rape seed, using several allocation principles. Sections 4 and 5 discuss and conclude.

8.2 ALLOCATION: PRACTICE, POLICY AND PROBLEMS

8.2.1 ALLOCATION METHODS

During the inventory phase of an LCA the problem of multifunctional processes, and thus of allocation, is often encountered. Following Guinée (2002), a multifunctional process is considered as "a unit process yielding more than one functional flow, i.e. co-production [more than one product outflow], combined waste processing [more than one waste inflow] and recycling [one or more product outflows and one or more waste inflows]".

Multifunctional processes are a problem for LCA because usually not all the functional flows are part of the same product system. Thus, a multifunctional process is part of the product system studied and also of other systems. The question is then, how to allocate the environmental impacts of this multifunctional process to the different product systems, i.e. to the different functional flows.

The LCA community has come up with various ways to address the multi-functionality problem. The on-going debate on allocation triggers the question whether there actually is a 'correct' way to address this problem. It can be argued that by focusing on the physical relationships behind the process this question can be answered positively. However, this argument has so far not been able to bring the allocation debate to an end (see also Weidema and Schmidt 2010 for a summary of recent discussions on allocation). Three types of reasons for this can be identified; (1) there are always various physical relationships to choose from for a multifunctional process, (2) different co-product can be expressed in different physical quantities (e.g. mass and energy), and (3) physical relationships do not necessarily reflect properly the ground for existence of a process (like mass for processes co-producing medicines in small amounts and fodder in big amounts).

In this article, the on-going debate on allocation is seen as a sign that the question above should be answered negatively. It follows in this respect the assertion of Guinée et al. (2004) that "the multi-functionality problem is an artefact of wishing to isolate one function out of many. As artefacts can only be cured in an artificial way, there is no 'correct' way of solving the multi-functionality problem, even not in theory." The most frequently used methods to solve this problem are shortly introduced below. The introduction discusses not only the rationales behind the methods, but also discusses their advantages and flaws:

- Subdivision: disentangling a process that has been recorded as a multi-functional unit process into the constituent mono-functional unit processes

- System expansion: avoiding the multi-functionality problem by broadening the system boundaries and introducing new processes and several functional units
- Physical partitioning: the artificial splitting up of a multifunctional process into a number of independently operating mono-functional processes, based on physical properties of the flows (e.g. mass, energy, carbon content, etc.)
- Economic partitioning: the artificial splitting is based on economic properties of the multifunctional process, such as the gross sales value or the expected economic gain

In order to come to a standardization of LCA, International Organization for Standardization (ISO) introduced a hierarchical approach for dealing with multi-functionality. The ISO 14044 allocation procedure (clause 4.3.4.2) prescribes subdivision or system expansion as a first step in order to avoid actual allocation. In case allocation cannot be avoided ISO prescribes physical partitioning as a second step. The procedure emphasizes that this type of partitioning should reflect the underlying physical relationship between the different products or functions. As a third step, when physical partitioning cannot be established, ISO prescribes allocation in a way that reflects another (e.g. economic) relationship between the different products or functions (ISO 14040 2006).

In addition to these allocation methods mentioned in the ISO standard, there is the often used approach of substitution:

- Substitution: the concept behind substitution is that the production of a co-product by the system studied causes another production process in another system to be avoided. This avoided production process results in avoided emissions, resource extractions etc. that should be subtracted from the studied product system

Several authors have argued that substitution is conceptually equivalent to system expansion (e.g. Ekvall and Tillman 1997; Finnveden and Lindfors 1998). Conceptually equivalent does not mean that system expansion and substitution provide the same results, but that they provide results that are compatible.[1] The two allocation methods share subsequently some advantages and disadvantages. Both methods, for example, increase the level of complexity by adding extra processes, either to be added, or to be subtracted. A consequence of the conceptual equivalency between the two approaches is that it is used as an implicit argument to choose for substitution, while still claiming compliance to ISO.

It is important to note, however, that there are also large differences between these methods. An important drawback that is particular for system expansion concerns the fact that the system provides more than one function, so that a multiple functional unit is used. It can be questioned whether an LCA that aims at studying the environmental burdens of one specific function, achieves this aim when it gets an answer for several functional units. Drawbacks specific for substitution are related to the various assumptions that have to be made. For example, it has to be argued which production process is actually avoided.

Physical partitioning is one of the simplest allocation methods to apply, and if one carefully chooses the physical characteristic used as basis for the method, it is quite straightforward to apply. However, determining the physical characteristics to be used as a basis for allocation can be challenging. Potentially relevant characteristics should relate to the purpose or use of the product. But co-products often have different purposes (or uses) and thus different characteristics may be relevant in understanding why they are sold. In many cases, the LCA practitioner can overcome this problem by selecting a physical characteristic that make sense for both product and co-product. However, such a common denominator cannot always be identified, e.g. a system that produces both meat and leather, or a waste incinerator that fulfills the function of waste processing and the function of energy production.

Economic partitioning is another often applied allocation method. By taking the economic value of different processes as a basis for allocation, economic partitioning addresses the economic motivation behind a multifunctional process. While some practitioners see this as strength, it can also be seen as the main drawback to economic partitioning. Another argument against economic partitioning is that prices can fluctuate independently from the long term economic value of a process. Also, the fact that prices can vary between different locations is sometimes seen as a disadvantage of economic partitioning (Ayer et al. 2007).

8.2.2 ALLOCATION IN POLICY GUIDELINES

Early political visions included high level of biofuel incorporations into transport fuels with no restrictions of origin or production pathways (CEC 2007). However, in a response to the more critical stances to bio-energy, governments have introduced directives with the intention to stimulate sustainable bio-energy only (RTFO 2007; Directive 2009/28/EC, 2009; LCFS 2007; EPA 2010). LCA plays an important role in these regulations as it often serves as the main tool to assess alternative energy production systems' reductions in GHG emissions.

The European Union and the USA have led the way in using LCA in regulatory schemes. The first schemes appeared in individual European Member States. The UK implemented the Renewable Transport Fuel Obligation (RTFO) in 2007, which requires transport fuel providers to report the sustainability level of the fuels provided in the UK (RTFO 2007). For the GHG criterion, the scheme required that reporting parties calculate the carbon intensity of their fuel based on a specified LCA methodology, including using a 'restricted' substitution method for allocation. The RTFO restricted the substitution method by only allowing some uses for specific co-products (e.g. rapeseed cake could only be used for animal fodder). However, for certain chains, it was not possible to identify the use of the co-product. In these cases, economic partitioning was used, as it was felt to be the closest allocation method to substitution (RTFO 2007; Table 1).

Having set ambitious targets for the use of biofuels in Europe (10 % renewable in the transport sector by 2020; CEC 2007), the European Commission published, in 2009, a directive with the goal to ensure the sustainability of biofuels and bioliquids (Directive 2009/28/EC, 2009). The Renewable Energy Directive (RED) defines a minimum threshold for GHG emission savings that must be achieved by bio-fuels to be considered renewable energy. The calculation methodology for GHG savings is also defined in the directive. The directive imposes to use energy content as

Table 1. Overview of bio-energy directives.

Legislation	Region covered	Allocation method
Renewable Transport Fuel Obligation	UK	Substitution whenever possible, if not allocation based on economic value
Renewable Energy Directive	European Union (all 27 Member States)	Allocation based on energy content except for electricity co-production for which it is substitution
Low Carbon Fuel Standard	California	Substitution whenever possible, if not allocation based on energy content
Renewable Fuel Standard 2010	USA	Substitution

basis for allocation, except for electricity that is co-produced with biofuel or bioliquid, and which, under certain conditions, should be allocated applying substitution (Directive 2009/28/EC 2009). As a European directive, the RED will be transposed into national legislation in European member states. In case of UK, this means that the guidelines of the RED are being implemented in the RTFO.

The US has also recently seen the development of two schemes regulating the GHG emissions of their transport fuel. The first, the Low Carbon Fuel Standard (LCFS 2007), was set in place in California. This scheme defines an average maximum carbon intensity target for the mix of transportation fuels used in California. Transport fuels that have lower carbon emissions than the target are awarded credits, which they can sell to compensate fuels that are too carbon intensive. The credits and debits are awarded based on the life cycle GHG emissions of transportation fuels. The LCFS requires substitution to be used as allocation method (LCFS 2007). However, in practice some chains use physical partitioning on the basis of energy content (CEPA 2009).

A federal regulation is also under preparation in the USA, the Renewable Fuels Standard 2010 (EPA 2010), which requires the

US Environmental Protection Agency (EPA) to calculate the carbon intensity of the biofuels becoming available most likely in the USA. EPA's results will then be used to classify the fuels into four different categories (cellulosic biofuel, biomass-based diesel, advanced biofuel and renewable fuel), which each have different volume targets. EPA performed their LCA calculations applying a substitution method in case of multi-functional processes.

8.2.3 DISCUSSION

The schemes presented in this section are not only distinguishable by their geographical scope. Their reporting requirements have led them to implement different allocation methods. Most European schemes require industries to calculate and report their GHG emissions themselves, so the allocation methods applied have to be simple. And indeed, even the RTFO only employs a 'restricted' substitution. In the American schemes, calculations are performed with support of given default values that industries have to use. These default values can only be changed under specific conditions, and only by the scheme's implementation body. Therefore, the American schemes can use somewhat more complex allocation methods.

This diversity in reporting requirements is confusing for bio-energy producers and users. As most modern markets, the bio-energy market has a global character and consists of international actors and relationships. Moreover, the use of different allocation methods in different schemes is not only confusing but also disturbing. After all, different allocation methods potentially lead to different assessments of a single bio-energy stream (Kim and Dale 2002; Wang et al. 2004; Guinée and Heijungs 2007; Thomassen et al. 2008; Bier et al. 2012). To assess whether it can be expected that the different requirements in the schemes above result in different assessments, they are applied on a case study below. To serve its purpose of illustration, a real life case study has been selected that is relatively simple and straightforward, focusing on GHGs only and leaving out of the analysis other impacts (including direct and indirect land use change that may obviously be an important issue) and further methodological discussions. In this way, the case serves as a suitable test on whether the problem is real or only hypothetical.

8.3 CASE STUDY ON RAPESEED

8.3.1 GOAL AND SCOPE

The discussion above shows that different countries promote different guidelines with respect to addressing multifunctional processes. The goal of this LCA study is to assess the influences of the choice of allocation on the outcomes of an LCA.

Previous studies have already used a similar approach using a hypothetical case (Guinée et al. 2009; Luo et al. 2009). In the present study, the allocation methods are applied on an existing bio-electricity chain. The chain selected is the rapeseed to bio-electricity chain.

The electricity production for the Dutch mix was used as a reference chain and renewable energy resources were not considered to contribute in this mix. This comparison took place only for one impact category: climate change.

The selected functional unit for the case study was: The production of 1 kWh low voltage electricity at the Dutch grid. For the case study, main data sources were Hamelinck et al. (2008) and Van der Voet et al. (2008). To check data from these sources and when additional data was needed, the ecoinvent database was consulted, especially for agriculture (Nemecek et al. 2000). The CMLCA 5.0 software, accessed via www.cmlca.eu, was used as a calculation platform.

8.3.2 LIFE CYCLE INVENTORY ANALYSIS

The chain consists of five main life cycle stages: the feedstock production, the feedstock transport, the conversion, the oil transport and the electricity generation. For the system boundaries definition, two main assumptions were made:

1. A distinction between 'negative' and 'positive' CO_2 emissions. All CO_2 emissions from the feedstock production phase (rapeseed cultivation) were considered (positive) emissions to the environment, while CO_2 fixation in the same phase was considered a negative emission (or extraction from the environment). For all other emission, this distinction was not made and the carbon emissions are still accounted for (including that released upon combustion).

2. Emissions from electricity production are included as well as emissions from the production of input materials and energy to all other processes (e.g. fertilizers production, electricity used for conversion processes, among the main ones).

The flow diagram in Fig. 1 provides an outline of all the major unit processes in the system. The flow diagram is based on the flow diagrams of Hamelinck et al. (2008) and of van der Voet et al. (2008). It is remodeled and specified for the present case study.

It was assumed that rapeseed is cultivated and produced in Northern Europe as well as it was assumed that the rapeseed straw generated during the harvesting process is plowed back to the ground replacing part of the nitrogen fertilizer. The rapeseed is then transported to the conversion plant where the oil will be extracted. The estimated requirement for this transportation is 150 tkm within the Netherlands or between Germany and the Netherlands.

Once the rapeseed is at the conversion plant, two main processes take place in order to extract the oil: (1) storage and (2) cold pressing of the rapeseed. Out of the pressing process, two products are obtained: the rapeseed oil and the rapeseed cake. This was the process for which the multi-functionality problem was solved by applying the different allocation methods. Hence, special focus is given to this process. Afterwards, the oil is transported from the conversion plant to the power plant where it is combusted in order to produce electricity. It was estimated again that transportation requirements was 150 tkm. Co-firing with heavy oil and

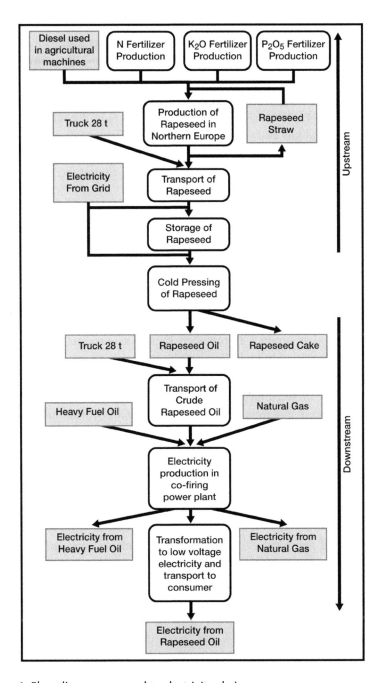

Figure 1. Flow diagram rapeseed to electricity chain

natural gas was the chosen technology in the chain for the bioelectricity generation process. This process involves also another multi-functionality problem due to the three economic inflows it has (heavy oil, natural gas and rapeseed oil). In order to concentrate on the multi-functionality problem from the conversion process (pressing process), the energy production was allocated on the basis of the energy content: 37 % to the rapeseed oil, 37 % to the natural gas and 46 % to the heavy oil. Two more processes take place in order to deliver the electricity to the consumer: the conversion of the electricity produced from high voltage to low voltage and the transportation of the electricity to the consumer.

Since the core point of the case study is to analyze the difference in the results when using different allocation methods, the allocation methods used in the bio-energy directives were applied to the multifunctional process in the conversion phase (pressing process). The allocation methods applied were thus: substitution, physical partitioning (on the basis of energy content) and economic partitioning (on the basis of proceeds).

As mentioned above, a difficulty for the substitution method is to determine which product is replaced by the co-product of the studied system. The case study includes two alternative cases of substitution: (1) substitution of soybean meal and (2) substitution of peas.

In the case of substitution of soybean meal, a loop is created: soybean meal is obtained together with soybean oil and soybean oil substitutes rapeseed oil. To deal with such a loop, practitioners can rely on two different approaches. First, ignore the fact that soybean meal and oil are coproduced and close the loop by including soybean meal alone. Second, extent the system by including the co-produced soybean oil and apply a form of partitioning. The use of the first approach implies a less realistic assumption, as these products are indeed co-produced. The use of the second approach is simpler and more realistic and still serves the illustrative purpose of the case study. Therefore, the second approach was chosen and economic partitioning was applied to the extraction process when substituting with soybean meal. It should be noted however that this is a simplification of the substitution method.

Therefore an alternative—and less realistic—case of substitution of peas has been added. This application is straightforward as peas production is not associated with any co-products requiring allocation.

The substitution in terms of protein can be considered as a simplification but it serves the illustrative purpose of the case study and is in line with energy policies.

The resulting substitution ratios are shown in Table 2. They are calculated based on the protein content of rapeseed cake, soybean meal and peas (Brookes 2001; Corbett 2008). Table 3 shows the allocation ratios used for the partitioning methods. With the allocation ratios, it is possible to allocate the burdens (emissions to air) between the rapeseed oil and rapeseed cake from the pressing process.

The emissions generated in processes taking place before the pressing process (i.e. upstream of the multifunctional process), are the ones allocated to the two products that result from pressing rapeseed (i.e. rapeseed oil and rapeseed cake) with the different allocation ratios from different allocation methods. The downstream emissions (those being emitted in processes after the pressing process) correspond to the total emissions of the chain calculated with a surplus method and subtracting the upstream emissions. The upstream and downstream emissions are shown respectively in Tables 4 and 5.

8.3.3 LIFE CYCLE IMPACT ASSESSMENT OF THE CASE STUDY

The only impact category analyzed was the climate change category. The inventory results of GHG emissions to air in kg were transformed to kilogram of CO_2 equivalents. The Global Warming Potential for a 100-year time horizon developed by the Intergovernmental Panel on Climate Change was used as the characterization factor (IPCC 2007).

The climate change profile obtained for the rapeseed oil to bioelectricity chain by using different allocation methods is shown in Table 5. The results are compared to a reference chain and the improvement is also calculated. The reference chain is the Dutch production mix based on fossil fuels and nuclear energy (Van der Voet et al. 2008). The composition of the Dutch electricity mix is given in Table 6 (Seebregts and Volkers 2005; CBS 2007):

The total GHG emissions of the reference chain are 0.715 Kg of CO_2 equivalents. The percentage of improvement is calculated by subtracting the total GHG emissions of the bio-electricity chain from those of the

Table 2. Substitution ratios.

Substituted product	Protein content (mass %)	Substitution ratio (rape/ substrate)
Soybean meal	45	0.75
Peas	24	1.5

Table 3. Allocation ratios used in case study.

Method	Rapeseed oil	Rapeseed cake
Economic partitioning	0.70	0.30
Physical partitioning	0.55	0.45

Table 4. Allocation ratios used in case study.

Elementary flow	Upstream emission to air (kg CO_2 eq)
Carbon dioxide to air	0.246
Nitrous oxide (N_2O) to air	0.00199
Methane to air	0.000324

Table 5. Partial inventory table for downstream part of bio-electricity chain.

Elementary flow	Downstream emission to air (kg CO_2 eq)
Carbon dioxide to air	0.012
Nitrous oxide (N_2O) to air	0.00
Methane to air	0.000019

reference chain and then dividing it by the total GHG emissions of the reference chain:

$$GHG_{reduction}(\%) = \frac{GHG_{emission,fossil\ chain} - GHG_{emission,bio-chain}}{GHG_{emission,fossil\ chain}} \times 100 \qquad (1)$$

As Table 7 shows, the allocation method used has a substantial influence on the results of the impact assessment. The method leading to the largest indicator of improvement is the substitution of peas (~60 %); followed by physical partitioning on energy basis (~35 %). Finally, the substitution of soybean meal and the economic partitioning lead to approximately 20 % improvement.

8.4 DISCUSSION

The outcomes of the case study show that the choice of allocation method can have a considerable impact on the outcomes of an LCA, even for a system that is small, and where the allocation issue has been restricted to one process. The outcomes in this study range between a 16 and 60 % improvement compared to a reference chain. And although substitution with peas tilts the picture with its extreme outcome, it should be noted that the other methods still produce outcomes that range from 16 to 33 %.

Allocation methods are frequently required in LCA, especially when complex systems, like energy production systems, are involved. At this moment, directives regarding the assessment of bio-energy production still prescribe different allocation methods. As the case study shows, this poses a problem because the outcomes of LCA differ strongly, depending on which directive is followed. In the EU, the Renewable Energy Directive is likely to lead to standardization of national arrangements but differences between the EU and the USA will remain. Also, transposition of the RED in national legislation might still result in diverse application of allocation methods due to differences in interpretation. And even when similar allocation methods are used within a single sector, different outcomes can be obtained due to methodological difficulties or a lack of reliable data.

As noted above, this diversity in policy directives is confusing and disturbing for bio-energy producers and bio-energy users. The resulting

Table 6. Reference chain: Dutch electricity mix.

Source	Mix (%)	Efficiency (%)	Remark
Natural gas	52.0	43	
Hard coal	43.6	39	
Nuclear	4.1		90 % pressure water reactor, 10 % boiling water reactor
Industrial gas	0.1	36	
Oil	0.1	44	

Table 7. Greenhouse gas emissions of bio-electricity chain.

Allocation method	GHG (in kg CO_2 eq)	Performance compared to reference chain	Improvement (%)
Economic partitioning (on the basis of proceeds)	0.604	−0.111	+16
Physical partitioning (on the basis of energy content)	0.477	−0.238	+33
Substitution (by soybean)	0.567	−0.148	+21
Substitution (by peas)	0.293	−0.422	+60

differences, due to methodological choices, in the assessments of countries are hard to justify in a policy context. Besides, this uncertainty adds on to other uncertainties for example those related to data issues. With the (economic) stakes high, the uncertainty due to methodological choices might lead to legal problems. Bio-energy producers, for example, may consider a different assessment in another country as an indirect trade barrier.

To avert such a situation, we argue that it is important to discriminate between analysis and policy related LCAs. In the history of LCA, important distinctions between LCA types have already been introduced, most notably attributional versus consequential LCAs, and recent attempts in the ILCD handbook (European Commission–Joint Research Centre–Institute for Environment and Sustainability 2010). Our distinction focuses not on type, but on requirements on LCAs. Analysis-related LCAs are LCA studies that are carried out for the purpose of understanding a certain system. They try to identify important impacts, main contributors to impacts, opportunities to reduce impacts or otherwise optimize the system, as well as to analyze the effects of data, assumptions, and choices. Understanding and presenting uncertainties and trade-offs in such an assessment adds to the aim of completeness. Policy-related LCAs, on the other hand, support the regulation of the production, trade and use of certain products. They try to support the governance of industrial systems through subsidizing or certifying desired products, or by taxing or banning undesired products. High levels of uncertainty in this context might lead to inconsistent policy, resulting in strategic behavior of involved actors or in legal disputes. We argue therefore that this difference in aims should be taken into account when setting up an LCA study. As the understanding of the system under study is the main aim of analysis-related LCA, trade-offs and uncertainties that are encountered during the performance of such LCAs can be handled in line with the views of the involved researcher as long as choices are transparently displayed. The main aim of policy related LCA is to deliver comparable results. As differences in the handling of trade-offs and uncertainties in LCAs can impede the comparability of results, it is of great importance to present clear and straight-forward applicable guidelines for such choices in a policy context.

We argue that there is not an objectively correct way to solve the multi-functionality problem, but the problem can be solved in a way that serves the aim of the LCA best. In a policy context, LCAs should contribute to long-term stability in the system, provide actors equal and full information, and create a level playing field. In other words, policy-related LCAs aim for consistency and robustness. This aim for robustness is not served by the existing guidelines of ISO. As discussed above, ISO strives in the first place for completeness. In practice, this turns out to be difficult due to methodological difficulties and problems with missing or unreliable data. The use of LCA in the policy context should therefore benefit to a great extent from a guideline based on robustness.

It is beyond the scope of this article to draw the outlines of such a guideline, but we foresee that the recommended allocation method within the bio-energy context will be physical partitioning based on energy content. After all, physical partitioning is relatively easy to apply, the data is unambiguous, the outcomes are stable over time and energy-content is the most common denominator of co-products in bio-energy LCAs. Although this choice will not be able to remove all uncertainty (it does not address for example data issues), the method's stability will increase the robustness of policy outcomes.

8.5 CONCLUSIONS

The aim of this article was to show to what extent a choice of allocation method can influence the outcomes of an LCA on bio-electricity production. The outcomes of a case study on a rapeseed-to-bio-electricity chain showed that variation between 16 and 60 % reduction of GHG emissions in comparison to a reference chain can be obtained depending on the allocation method applied. These findings emphasize the urgency to develop a clear guideline for LCA practice as using different allocation methods can, intentionally or unintentionally, result in very different outcomes.

Current policies, originating from different regions, prescribe different allocation methods. The recent EU's Renewable Energy Directive introduces some uniformity for EU member states but differences with the US and other world regions will remain. Moreover, national governments can still end up with different regulations due to different interpretations of the EU directive. The undesirability of this situation lays it the uncertainty for bio-energy producers and consumers.

To overcome this situation, we focused on an important difference between scientific and policy LCAs. Whereas the former aims for completeness, the latter aims for robustness. The use of LCA in the policy context will benefit largely from the acceptance of this difference and by drawing up a guideline that is based on the aim of robustness. This paper serves as a starting point for realizing such a guideline. We think that in such a guideline physical partitioning on energy content is the favored allocation method. However, we do not deny the fact that physical partitioning on energy content has its own drawbacks. We urge therefore that the drafting of this guideline should be accompanied by an on-going dialog between practitioners and commissioners to strengthen the use of LCA as a policy tool.

FOOTNOTES

1. In order to compare system I that produces products A and B simultaneously with system II that produces product B, it is the same to add to system II the production system of product B (system expansion) or to subtract from system I the same production system (substitution).

REFERENCES

1. Ayer NW, Tyedmers PH, Pelletier NL, Sonesson U, Scholz A (2007) Co-product allocation in life cycle assessments of seafood production systems: review of problems and strategies. Int J Life Cycle Assess 12(7):480–487
2. Bier JM, Verbeek CJR, Lay MC (2012) An eco-profile of thermoplastic protein derived from blood meal. Part 1: allocation issues. Int J Life Cycle Assess 17(2):208–219
3. Bindraban P, Pistorius R (2008) Biofuels and food security. Plant Research International, Wageningen
4. Brookes G (2001) The EU animal feed sector: protein ingredient use and the implications of the ban on use of meat and bonemeal. BrookesWest, Elham, Canterbury, UK
5. CBS (2007) Compendium voor de Leefomgeving. Available from: http://www.compendiumvoordeleefomgeving.nl/onderwerpen/nl0006-Energie-en-milieu.html?i=6. Accessed on 5 June 2011.
6. CEC (Commission of the European Communities) (2007) Biofuels Progress Report: Report on the progress made in the use of biofuels and other renewable fuels in the member states of the European Union. Brussels, Belgium
7. Chum H, Faaij APC, Moreira JR, Berndes G, Dhamija P, Dong H, Gabriella B, Goss Eng A, Lucht W, Mapako M, Masera Cerutte M, McIntyre T, Minowa T, Pingoud K (2011) Bioenergy. In: Edenhoffer O, Pichs-Madruga R, Sokona Y (eds) IPCC Special Report on Renewable Energy Sources and Climate Change. Cambridge University Press, Unitid Kingdom and New York, NY, USA

8. Corbett RR (2008) Peas as a protein and energy source for ruminants. Available from http://www.wcds.ca/proc/1997/ch18-97.htm. Accessed on 5 June 2011

9. De Fraiture C, Giordano M, Liao Y (2008) Biofuels and implications for agricultural water use: blue impacts of green energy. Water Pol 10(1):67–81

10. Directive 2009/28/EC (2009) Directive 2009/28/EC of the European Parliament and of the Council of April 23, 2009 on the promotion of the use of energy from renewable sources and amending and subsequently repealing Directives 2001/77/EC and 2003/30/EC. Brussels, Belgium

11. Ekvall T, Tillman A (1997) Open-loop recycling: criteria for allocation procedures. Int J Life Cycle Assess 2(3):155–162

12. CEPA (California Environmental Protection Agency) (2009) Detailed California-Modified GREET Pathway for Conversion of Midwest Soybeans to Biodiesel (Fatty Acid Methyl Esters-FAME)

13. EPA (United States Environmental Protection Agency) (2010) RFS2: Regulation of Fuel and Fuel Additives: Modifications to Renewable Fuel Standard Program

14. European Commission–Joint Research Centre–Institute for Environment and Sustainability (2010) International Reference Life Cycle Data System (ILCD) Handbook–general guide for life cycle assessment—detailed guidance. Luxemburg

15. Finnveden G, Lindfors L (1998) Data quality of life cycle inventory data—rules of thumb. Int J Life Cycle Assess 3(2):65–66

16. Guinée J (ed) (2002) Handbook on life cycle assessment operational guide to the ISO standards. Kluwer, Dordrecht

17. Guinée JB, Heijungs R (2007) Calculating the influence of alternative allocation scenarios in fossil fuel chains. Int J Life Cycle Assess 12(3):173–180

18. Guinée J, Heijungs R, Huppes G (2004) Economic allocation: examples and derived decision tree. Int J Life Cycle Assess 9(1):23–33

19. Guinée J, Heijungs R, van der Voet E (2009) A greenhouse gas indicator for bioenergy: some theoretical issues with practical implications. Int J Life Cycle Assess 14(4):328–339

20. Hamelinck C, Koop K, Kroezen H, Koper M, Kampman B, Bergsma G (2008) Technical specification: greenhouse gas calculator for biofuels, Ecofys/CE Utrecht

21. IPCC et al (2007) In: Solomon S, Qin D, Manning M (eds) Climate change 2007: The physical science basis. Contribution of Working Group I to the fourth assessment report of the Intergovernmental Panel on Climate Change. Cambridge University Press, New York

22. ISO 14040 (2006) International Standard ISO 14040: environmental management–life cycle assessment—principles and framework Geneva, Switzerland

23. Kim S, Dale BE (2002) Allocation procedure in ethanol production system from corn grain I. system expansion. Int J Life Cycle Assess 7(4):237–243

24. LCFS (Low Carbon Fuel Standard) (2007) Low Carbon Fuel Standard. California

25. Luo L, van der Voet E, Huppes G, Udo de Haes H (2009) Allocation issues in LCA methodology: a case study of corn stover-based fuel ethanol. Int J Life Cycle Assess 14(6):529–539

26. Nemecek T, Heil A, Huguenin O, Meier S, Erzinger S, Blaser S, Dux D, Zimmermann A (2000) Life cycle inventories of agricultural production systems Final Report Ecoinvent. Swiss Centre for Life Cycle Inventories, Dubendorf

27. Reijnders L, Huijbregts M (2008) Palm oil and the emission of carbon-based greenhouse gases. J Clean Prod 16(4):477–482

28. Searchinger T, Heimlich R, Houghton R, Dong F, Elobeid A, Fabiosa J, Tokgoz S, Hayes D, Yu T (2008) Use of US croplands for biofuels increases greenhouse gases through emissions from land-use change. Science 319(5867):1238

29. Seebregts A, Volkers C (2005) Monitoring Nederlandse elektriciteitscentrales 2000–2004 ECN Beleidsstudies. Petten, the Netherlands, ECN

30. SenterNovem (2008) The greenhouse gas calculation methodology for biomass-based electricity, heat and fuels. Utrecht

31. Thomassen MA, Dalgaard R, Heijungs R, de Boer I (2008) Attributional and consequential LCA of milk production. Int J Life Cycle Assess 13(4):339–349

32. RTFO (Renewable Transport Fuel Obligations) (2007) Renewable Transport Fuel Obligations Order 2007

33. UNEP (United Nations Environmental Program) (2009) Guidelines for social life cycle assessment of products. Paris, France

34. United States Department of Energy (2010) Growing America's energy future: renewable bioenergy

35. Van der Voet E, Van Oers L, Davis C, Nelis R, Cok B, Heijungs R, Chappin E, Guinée J (2008) Greenhouse gas calculator for electricity and heat from biomass. CML Institute of Environmental Sciences, Leiden

36. Van der Voet E, Lifset RJ, Luo L (2010) Life-cycle assessment of biofuels, convergence and divergence. Biofuels 1(3):435–449

37. Wang M, Lee H, Molburg J (2004) Allocation of energy use in petroleum refineries to petroleum products. Int J Life Cycle Assess 9(1):34–44

38. Weidema BP, Schmidt JH (2010) Avoiding allocation in life cycle assessment revisited. J Ind Ecol 14(2):192–195

39. Worldwatch Institute (2007) Biofuels for transportation: global potential and implications for sustainable agriculture and energy in the 21st century. Washington DC

CHAPTER 9

An Improved Optimal Capacity Ratio Design Method for WSB/HPS System Based on Complementary Characteristics of Wind and Solar

XIAOJU YIN, FENGGE ZHANG, ZHENHE JU, AND YONGGANG JIAO

9.1 INTRODUCTION

Separate photovoltaic or wind power generation can achieve a major output fluctuation due to random and intermittent wind/solar resources and needs prodigious battery volume to satisfy load. Actually, the photovoltaic and wind power generation have commendable complementary characteristics in time distribution. The solar energy is adequate and the wind is weak in daytime, while the solar is weak and the wind is strengthened due to the earth's surface difference in temperature at night. In summer, the solar is strong but wind is weak, while the wind is strong but the solar is weak in winter. Taking full advantage of the complementary characteristics of wind and solar can achieve a high power supply reliably, while it can reduce the total cost of the system in the meantime [1].

© 2015 Xiaoju Yin et al. Mathematical Problems in Engineering, *Volume 2015 (2015), Article ID 703623. doi: http://dx.doi.org/10.1155/2015/703623. Creative Commons Attribution License (http://creativecommons.org/licenses/by/3.0/).*

There are many optimal design methods target-based for stand-alone WSB hybrid power system. The restraint conditions mainly include power supply reliability, total cost of system, and environmental protection. Actually, the traditional optimal design method considers less restraint conditions which include current, frequency, rate in charge/discharge cycles of the battery, and system backup capacity. Moreover, the complementary characteristics of wind and solar resources are not considered a restraint condition [2–8]. Therefore, this paper proposed an improved optimal design method for stand-alone wind-solar-battery hybrid power system. First, the paper formulated WSB/HPS optimal sizing principle. Then, an optimal sizing mode was established and an optimal ratio strategy was proposed. Finally, an average power of 100 kW system for an example is taken to simulate and verify the improved optimal design method for stand-alone wind-solar-battery hybrid power system.

9.2 PRELIMINARIES

9.2.1 STRUCTURE AND WORKING PRINCIPLE OF WIND TURBINE

Wind turbines consist of wind turbine and generator [9]. In the small wind generator, wind power working process is shown in Figure 1. The kinetic energy of airflow on the wind wheel is converted to mechanical energy by the aerodynamic rife wheel rotation which promotes the wind wheel spin. Since small wind turbine power agencies do not consume energy usually, therefore, the wind turbine generators are driven directly. Eventually, the derived mechanical energy is transformed into electrical energy to supply for the load.

Wind energy is the kinetic energy of the air and therefore is determined by the wind speed and air density, which is given by the following equation:

$$E = \frac{mv^2}{2} \tag{1}$$

where E is the kinetic energy of air, m is the quality of the object, and V is the movement speed of the object.

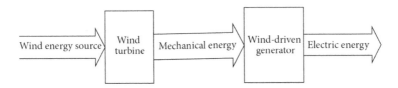

Figure 1. Flow diagram rapeseed to electricity chain

Obviously, the above equation can be written as

$$E = \frac{(pvA)v^2}{2} = \frac{pAv^3}{2}$$ (2)

where E is the air density and m is the wind swept area.

Since all the energy of air almost is composed of the kinetic energy, therefore, the above equation is expressed by the power form:

$$P = \frac{pAv^3}{2}$$ (3)

In fact, all of the wind turbines are not able to convert wind energy into mechanical energy, so the actual power of the wind wheel is

$$P = 0.5pv^3 AC_p$$ (4)

where C_p is wind energy utilization factor, whose theoretical maximumis 0.593.That is, the maximumpower obtained can be 59.3% of the wind energy of the swept area range.

As is well known, wind power requires AC to DC rectifier section. In the rectification process, the full bridge rectifier technology of uncontrollable diode is employed in small power generation systems. Since the performance characteristics of a large loss are not obvious, its low cost is considered in this paper without affecting the power generation efficiency.

Figure 2 is a waveform diagram of the output voltage of the rectifier front of wind turbines. It will output stable DC voltage after the full-bridge rectifier filters.

9.2.2 THE WORKING PRINCIPLE AND CHARACTERISTICS OF SOLAR MODULES

First, the power generation principle and equivalent circuit components of solar modules are introduced.

The physical mechanism of the solar module is very similar to the classic PN junction diode. The solar module can use the photovoltaic effect to convert solar energy into electrical energy, and its power generation principle is shown in Figure 3. When solar module receives solar radiation, PN junction carriers will generate an electromotive force gradient that is accelerated in electric field. Once there is a load access, the circuit will occur. The unabsorbed photons will cause the battery temperature and thus emit into the environment.

Under standard illumination conditions, the rated output voltage is 0.48V. So multiple solar cells need to be connected together to obtain a higher output voltage and a large power capacity. Figure 3 is an equivalent circuit of solar modules.

When the lighting conditions are unchanged, photocurrent I_{ph} is a constant. It can be considered a constant current source. Output current I is equal to the photocurrent I_{ph} minus the diode current I_d and the leakage current I_{sh} is derived. The series resistance R_s of the current source resistance is determined by the PN junction depth, contact resistance, and impurity.

The diode current is given by the classical diode current expression:

$$I_d = I_0 \left\{ \exp\left[\frac{q(V + IR_s)}{AKT}\right] - 1 \right\} \qquad (5)$$

where $q = 1.6 \times 10^{-19}$ C is the electronic charge, $K = 1.38 \times 10^{-23}$ J/K is Boltzmann's constant, I_0 is the reverse saturation current of diode, A is the ideality factor of PN junction, and T is the absolute temperature.

Therefore, the output-side current is

$$I = I_{ph} - I_0 \left\{ \exp\left[\frac{q(V + IR_s)}{AKT} - 1\right] \right\} - \frac{V + IR_s}{R_{sh}} \qquad (6)$$

In the actual solar cells, the leakage current is very small and is therefore usually negligible.

In the following, the operating characteristics of solar modules are introduced.

Based on the above derivation of the mathematical model of solar modules, the photovoltaic cells model is set up using the Matlab Simulink modeling modules as shown in Figure 4.

According to the simulation model, the solar cell output characteristic curve is plotted where the light intensity is different (the outside temperature is 25°C), as is shown in Figure 5. It can be seen from the figure that the output power of solarmodules with an output voltage U and current I is nonlinear and there is a unique point of maximum power output. Our system is designed to track the position of the maximum power point of the solar module to keep the maximum output power, whose aim is to increase the utilization of solar energy.

For the establishment of a stable and balanced voltage to achieve full time power supply system, the energy storage units must be included in the configuration around wind turbine, solar modules, and load. That is, they attend to store power of wind turbines and solar modules and supply stable power to the load. In practical engineering, as the most economical and convenient storage methods, lead-acid batteries are widely used in small and medium sized wind and solar power generation system. Hybrid energy storage system structure is shown in Figure 6. In wind and solar power generation system, the controller is switched on and off to control the super capacitors and batteries charge and discharge strategy, to maximize the advantages of the above two storage forms, and to improve the stability and performance of the power storage unit in the system.

9.3 CAPACITY OPTIMIZATION RATIO PRINCIPLES FOR WSB/HPS SYSTEM

In the considered WSB/HPS system, the photovoltaic power generation, wind power generation, and battery assemble are placed on DC bus. The structure of the stand-alone WSB/HPS is proposed in Figure 6.

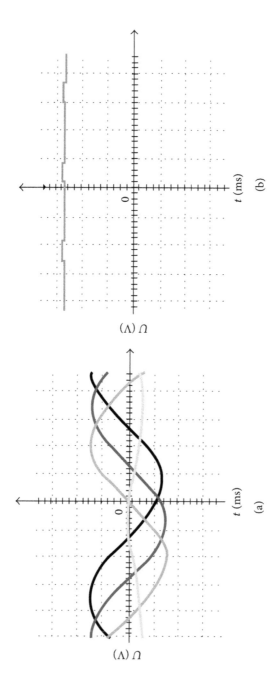

Figure 2. Generated voltage curve of wind turbine. (a) Three-phase alternating current generated by the wind power generator. (b) The full-bridge rectifier filtered voltage waveform.

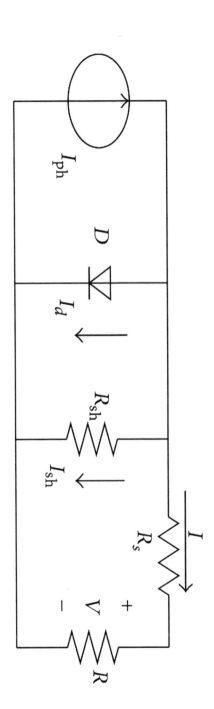

Figure 3. Solar photovoltaic battery equivalent circuit.

When the efficiency of wind power and photovoltaic output is low, the batteries are in the course of discharging. Then, the system power relationship is given as

$$P_L(t) = P_G(t) + P_{pv}(t) - P_{bdch}(t) \tag{7}$$

where $P_G(t)$ and $P_{pv}(t)$ are photovoltaic and wind power generation unit output, respectively, $P_{bdch}(t)$ is battery discharging power, and $P_L(t)$ is load power. If the system could not satisfy load power supply, a part of load must be switched off or backup power source must be launched. When photovoltaic and wind power are sufficient for load, the batteries are charged by redundant power. So the power relation is given by

$$P_L(t) = P_G(t) + P_{pv}(t) - P_{bdch}(t) \tag{8}$$

If the battery is fully charged, there is excess power and then the solar and wind power quit MPPT operation. It canbe seen from (7) and (8) that the considered WSB/HPS system can make the best complementary characteristics by optimizing wind/PV capacity ratio and reduce the capacity and depth of discharge and charge/discharge cycles of the battery. In summary, the optimization target of WSB/HPS includes a high power supply reliability and minimum cost by taking full advantage of complementary characteristics of wind and solar.

9.4 CAPACITY OPTIMIZATION RATIO MODEL

The reliability loss of power supply probability (LPSP) signifies the reliability level of power supply, which is defined as

$$f_{LPSP} = \frac{\sum_{i=1}^{N}[P_L(t_i) - (P_G(t_i) + P_{pv}(t_i) + P_{bdch}(t_i))]}{\sum_{i=1}^{N} P_L(t_i)} \tag{9}$$

where $P_{pv}(t_i)$, $P_G(t_i)$, $P_{bdch}(t_i)$, and $P_L(t_i)$ are PV, wind, battery, and load power at "t_i" time, respectively. N is the number of sampling interval points. Obviously, the less the f_{LPSP}, the higher the power supply reliability [4]. The summation of photovoltaic and wind generation output power is relative to load power fluctuation. D_L signifies the complementary characteristics of solar and wind generation, which is defined as

$$D_L = \frac{1}{\overline{P}} \sqrt{\frac{1}{N} \sum_{i=1}^{N} (P_L(t_i) + P_G(t_i) - P_L(t_i))^2}$$

(10)

where $\overline{P}$ is the average load power. Obviously, the D_L is less when the curve of the complementary characteristics of solar and wind generation is nearer to load power curve. Meantime, the capacity and depth of discharge and charge/discharge cycles of battery are reduced [5]. Therefore, the complementary characteristics of solar/wind generation are better. Initial investment cost of distributed generator (DG) is presented as

$$C_C = (N_{pv}C_{pv} + N_G C_G + N_b C_b)f_{cr}$$

(11)

where C_{pv}, C_G, and C_b are photovoltaic panel unit-price, wind turbine unit-price, and battery unit-price, respectively. N_{pv}, N_G, and N_b are the number of photovoltaic panels, the number of wind turbines, and the number of batteries, respectively. f_{cr} is depreciation coefficient, which is defined as

$$f_{cr} = \frac{r(1+r)^{L_f}}{(1+r)^{L_f} - 1}$$

(12)

where "r" is the depreciation rate; L_f is the project life time. Then, the operation and maintenance costs of the considered WSB/HPS system are presented as

$$C_{OM} = \sum (C_{pv}^{OM} t_{pv} + C_G^{OM} t_G + C_b^{OM} t_b)$$

(13)

where C_{pv}^{OM}, C_G^{OM}, and C_b^{OM} are PV power, wind power, battery charge and discharge operation, and maintenance costs under unit time, respectively. t_{pv}, t_G, and t_b are the running time of photovoltaic cells, wind generation unit, and battery unit, respectively. The replacement cost of unit DG and the total cost ofWSB-HPS system are given by

$$C_R = C_{pv}^R + C_G^R + C_b^R$$
$$C_A = C_C + C_{OM} + C_R \tag{14}$$

where C_{pv}^R, C_G^R, and C_b^R are the number of photovoltaic panels, wind turbines, and batteries, respectively.

The following constraint conditions are considered in this paper.

9.4.1 DG UNIT QUANTITY CONSTRAINT.

DG unit installation ground area is regarded as maximum installed capacity constraint. In general, the distance between two adjacent wind generators is 6–10 times of diameter length in the prevailing wind direction. Being perpendicular to the prevailing wind direction, the distance is 3–5 times of diameter length. Meanwhile, the line distance is 8 times of diameter length and column distance is 4 times of diameter length. Therefore, the quantity of wind generators should satisfy (15) and the quantity of installation PV and battery should satisfy (16) and (17):

$$N_G \leq \left[\frac{L}{8d} + 1\right] \cdot \left[\frac{W}{4d} + 1\right] \tag{15}$$

$$N_{pv} \leq \left[\frac{S}{S_{pv}}\right] a_{pv} \tag{16}$$

$$N_b \leq \left[\frac{S}{S_b}\right] \tag{17}$$

where S_{pv} and S_b are PV and battery area, respectively, and a_{pv} is shading coefficient.

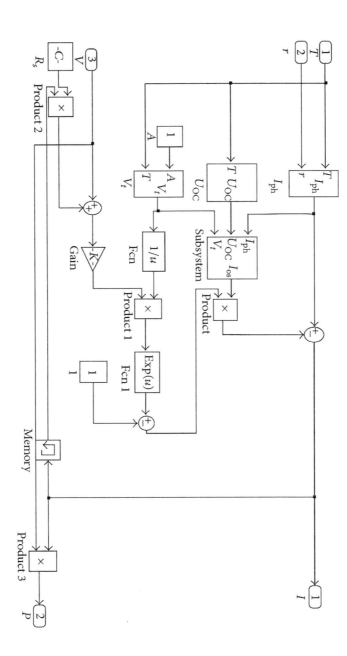

Figure 4. The photovoltaic cell model based on Simulink.

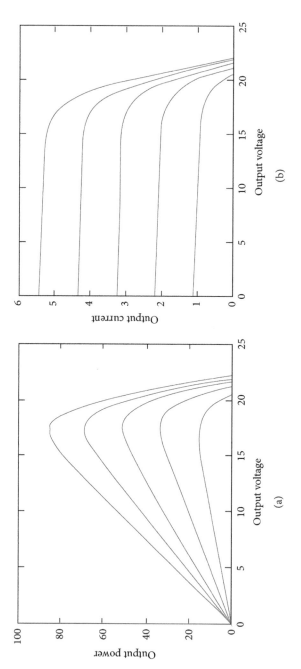

Figure 5. The output characteristic curve of solar cell under different light intensity. (a) Solar cell P-V characteristic curves. (b) Solar cell I-V characteristic curves.

9.4.2 DG UNIT MINIMUM POWER CONSTRAINT

Considering that the photovoltaic generation power output is zero at night, wind generation provides power supply for load. So wind generation should provide average load power at least. In the same way, if wind source is insufficient at daytime, photovoltaic generation should also provide average load power at least. If both the light and wind sources are insufficient, battery must provide load power. Then, an assumption is given below. The battery should ensure that the number of load sustaining running days is λ. Therefore, the minimum quantity of wind, PV, and battery unit installationmust satisfy the following equation:

$$N_G \geq \frac{\overline{P_{Ld}}}{P_G}$$

$$N_{pv} \geq \frac{\overline{P_{Ld}}}{P_{pv}} \tag{18}$$

$$N_b \geq \frac{\lambda W_{Ld}}{\eta C_b U_b D_{ODmax}}$$

where $\overline{P_{Ld}}$ is daily average load power and P_G and P_{pv} are each wind generator and each photovoltaic output power, respectively. W_{Ld} is daily load capacity (kW•h). C_b and U_b are capacity and voltage of unit battery, respectively. D_{ODmax} is maximum depth of discharge. η is battery discharge ratio.

9.4.3 SYSTEM BACKUP CAPACITY CONSTRAINT

Taking the increasing load or faulty units in the system into account, it is necessary that there is enough backup capacity in this system. Consequently, the overall maximum output power of DG units must ensure that there is incremental $m\%$ of load supply power. That is, the following inequality should be satisfied:

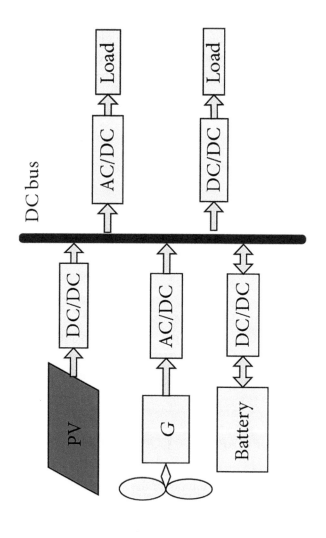

Figure 6. Structure of the stand-aloneWSBHPS.

$$\sum P_k \geq (1 + m\%)P_L \qquad (19)$$

where k denotes photovoltaic wind and/or battery unit.

9.4.4 BATTERY CHARGE/DISCHARGE CONSTRAINT

This paper established the mathematical representation of battery based on a class of dynamic charge/discharge models named KiBaM model. Thismodel can forecast battery capacity and port realtime voltage. Taking battery lifetime into consideration, the charge/discharge course of battery could be strictly restricted. The state of charge should satisfy (20). The charge ratio r_{ch} and discharge ratio r_{dch} should satisfy (21), where r_{ch-R} and r_{dh-R} are setpoints.The charge/discharge currents of batteries I_{ch} and I_{dh} cannot exceed their corresponding maximum values I_{chmax} and I_{dhmax}; that is, (22) should be satisfied. The charge/discharge power of battery (P_{bch} and P_{bdch}) should satisfy (23), where the value of P_{bchmax} and $P_{bdchmax}$ can be achieved based on KiBaM model:

$$S_{OCmin} \leq S_{OC} \leq S_{OCmax} \qquad (20)$$

$$r_{ch} \leq r_{ch\_R}, \qquad r_{dh} \leq r_{dh\_R} \qquad (21)$$

$$I_{ch} \leq I_{chmax}, \qquad I_{dh} \leq I_{dhmax} \qquad (22)$$

$$0 \leq P_{bch} \leq P_{bchmax}$$
$$0 \leq P_{bdch} \leq P_{bdchmax} \qquad (23)$$

9.4.5 BATTERY CHARGE/DISCHARGE FREQUENCY CONSTRAINT

As it is well known, the charge/discharge frequency influences battery's lifetime prodigiously. Therefore, the charge/discharge times and depth

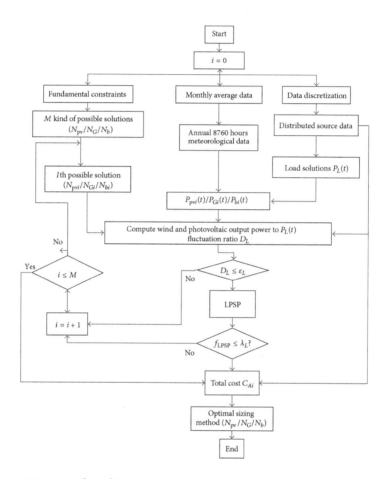

Figure 7. Program flow diagram of the optimal configuration.

of discharge should be limited. The constraint of discharge depth is determined by (20). In a scheduling period, the times of battery charge and discharge cycle N_C should not exceed the limit value N_{Cmax}. The value of N_{Cmax} is integratally decided by the load prediction, the importance of the load, and thebattery life androle inthe running system. That is, (24) should be satisfied. Since the scheduling period, S_{OC}, may be different, it is difficult to quantify the number of charge and discharge cycles of battery. This paper utilizes the equivalent number of charge/discharge cycles and life curve of battery to calculate the value of N_C,

$$N_C \leq N_{Cmax} \tag{24}$$

9.4.6 POWER SUPPLY RELIABILITY CONSTRAINTS

As mentioned above, the reliability loss of power supply probability (LPSP) f_{LPSP} signifies power supply reliability. Typically, the load power shortage should be in an acceptable range; that is, λ_L is permissible in the following inequality:

$$f_{LPSP} \leq \lambda_L \tag{25}$$

9.4.7 COMPLEMENTARY CHARACTERISTICS OF PHOTOVOLTAIC AND WIND POWER GENERATOR CONSTRAINTS

To make full use of wind and solar characteristics of the system, the index D_L must be less than the reference value ε_L,

$$D_L \leq \varepsilon_L \tag{26}$$

9.5 INDEPENDENT RUNTIME OPTIMIZATION STRATEGY FOR WSB/HPS SYSTEM

The total minimum cost of the considered WSB/HPS is chosen as optimization objective function in this paper. The relevant expression is presented in (27). The corresponding constraint conditions are given by (15)–(26),

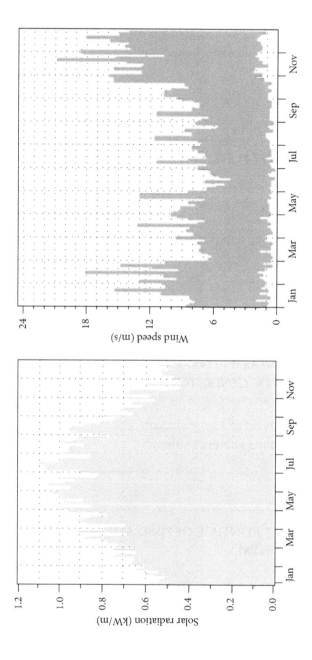

Figure 8. The PV Illumination curve and wind speed curve.

$$minf = minC_A \qquad (27)$$

According to the maximum of DG unit capacity constraints (15)–(17) and the minimum of DG unit power constraint (18), the ranges of N_G, N_{pv}, and N_b can be determined. Then, considering the system backup capacity constraint (15), the various possible capacity ratios of wind/photovoltaic/battery can be determined. According to wind/photovoltaic/battery output models (12), (13), and (15) and the battery charge and discharge constraints (20)–(24), the output power of wind power, photovoltaic, and battery can be calculated. Consequently, the values of f_{LPSP} and D_L could be figured out. An improved optimization scheme of wind/solar/battery capacity ratio is proposed under the reliability constraint (25) and the complementary characteristics constraint (26). A program flow diagram of the optimal configuration is proposed as shown in Figure 7.

9.6 EXAMPLE ANALYSIS

9.6.1 THE EXAMPLE AND CALCULATIONS

In order to verify that the proposed WSB-HPS capacity ratio optimization design method is reasonable and predominant, this paper presents an example to illustrate how to design an optimal capacity ratio for WSB/HPS system. The simulation software HOMER is employed in this paper [6, 7].

The relevant parameters are assumed below. The average power is 100 kW. The peak power is 315 kW. The load rate is 33%. The maximum load utilization hours are 2976 h. The day variation range is ±10% of the load as WSB-HPS related to the local load. Figure 8 shows the wind speed and the annual light curves which are obtained based on monthly average weather data after discretizing operation. After calculating the annual declination angle of the sun every day, the elevation angle and azimuth of the sun every hour, the corresponding discretizing illumination, and the annual hourly photovoltaic panel radiation are mainly acquired based on empirical formula. The annual discretized meteorological data reflect the basic characteristics of the weather in this example. Taking 100 kW as the average power of the load, we choose wind turbines of 35 kW rated power. Taking the PV array port voltage and power levels of the system into account, we choose a kind with single maximum power of 200 W and corresponding voltage of 24.5 V. When the WSB/HPS system is running, the process of battery charge and discharge is frequent, and therefore, the deep cycle VRLA batteries with monomer capacity of 600 Ah are chosen.

The traditional optimization program can be implemented independently using the HOMER software. The proposed improved optimization program in this paper can realize the calculation of f_{LPSP} and C_A by HOMER software. Consequently, the optimal wind/photovoltaic/battery capacity ratio can be exported to meet all abovementioned constraints. It should be pointed out that the calculation procedure assumes that wind turbines and photovoltaic units are running in MPPT state.

9.6.2 OPTIMIZATION CONFIGURATION RESULTS FOR WSB-HPS SYSTEM

Table 1 shows the comparison of configuration results between the traditional and improved optimization schemes. Figure 9 shows the output power curves of wind power generation and photovoltaic generation. The fluctuation of output power is reduced because a restraint condition that takes full advantage of the complementary characteristics of wind and solar is added. Consequently, fewer batteries are needed in the system, which can ensure power supply reliability and reduce the depth of discharge and charge/discharge cycles to extend lifetime. The cost of whole system slightly rises in that the price of photovoltaic cell panel is higher.

9.7 CONCLUSIONS

An improved optimal capacity ratio design method based on complementary characteristics of wind and solar is proposed in this paper. Under the precondition that the reliability of system is guaranteed, the total minimum cost of the WSB/HPS system is achieved by taking full advantage of wind and solar complementary characteristics. Since all the constraints

Table 1. Comparison of configuration results between the traditional and improved optimization schemes.

Optimization scheme	Photovoltaic cell panel	Wind-driven generator	Number of batteries	Total cost ($)
Traditional scheme	2200	10	1900	1327060
Improved scheme	2300	11	1630	1336532
Optimization scheme	Unit power cost $/(kWh)	D_L		f_{LPSP}
Traditional scheme	0.133	1.36		0.1
Improved scheme	0.134	1.25		0.09

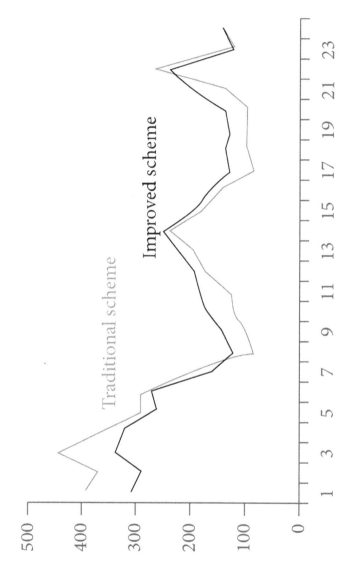

Figure 9. The PV Illumination curve and wind speed curve.

in the energy conversion process of system are almost proposed in the optimization model, such as the current, depth, and times constraints of battery charge and discharge, the quantity constraint of DG units, and so forth, the yielding optimal results are more accurate and reasonable.

REFERENCES

1. T. Esram and P. L. Chapman, "Comparison of photovoltaic array maximum power point tracking techniques," IEEE Transactions on Energy Conversion, vol. 22, no. 2, pp. 439–449, 2007. View at Publisher · View at Google Scholar · View at Scopus

2. R. Chedid and S. Rahman, "Unit sizing and control of hybrid wind-solar power systems," IEEE Transactions on Energy Conversion, vol. 12, no. 1, pp. 79–85, 1997. View at Publisher · View at Google Scholar · View at Scopus

3. A. Woyte, V. V. Thong, R. Belmans, and J. Nijs, "Voltage fluctuations on distribution level introduced by photovoltaic systems," IEEE Transactions on Energy Conversion, vol. 21, no. 1, pp. 202–209, 2006. View at Publisher · View at Google Scholar · View at Scopus

4. B. S. Borowy and Z. M. Salameh, "Methodology for optimally sizing the combination of a battery bank and PV array in a Wind/PV hybrid system," IEEE Transactions on Energy Conversion, vol. 11, no. 2, pp. 367–373, 1996. View at Publisher · View at Google Scholar · View at Scopus

5. G. B. Shrestha and L. Goel, "A study on optimal sizing of stand-alone photovoltaic stations," IEEE Transactions on Energy Conversion, vol. 13, no. 4, pp. 373–378, 1998. View at Publisher · View at Google Scholar · View at Scopus

6. P. Lilienthal, T. Lambert, and T. Ferguson, "Hybrid optimization model for electric renewable (HOMER) [EB/OL]," National Renewable Energy Lab (NREL) 2000-02-14 [2010-11-08], http://www.nrel.gov/HOMER.

7. J. H. Aylor, A. Thieme, and B. W. Johnson, "A battery state-of-charge indicator for electric wheelchairs," IEEE Transactions on Industrial Electronics, vol. 39, no. 5, pp. 398–409, 1992. View at Publisher · View at Google Scholar · View at Scopus

8. V. Mummadi, U. Katsumi, and U. Katsumi, "Feedforward maximum power point tracking of PV systems using fuzzy controller," IEEE Transactions on Aerospace and Electronic Systems, vol. 38, no. 3, pp. 969–981, 2002. View at Publisher · View at Google Scholar · View at Scopus

9. K. H. Chao, C. J. Li, and M. H. Wang, "A maximum power point tracking method based on extension neural network for PV systems," in Advances in Neural Networks—ISNN 2009, vol. 5551 of Lecture Notes in Computer Science, pp. 745–755, Springer, Berlin, Germany, 2009.

CHAPTER 10

Impact of the Choice of Emission Metric on Greenhouse Gas Abatement and Costs

MAARTEN VAN DEN BERG, ANDRIES F. HOF, JASPER VAN VLIET, AND DETLEF P VAN VUUREN

10.1 INTRODUCTION

While carbon dioxide (CO_2) has clearly the largest contribution to anthropogenic climate change several other gases also play a significant role, including methane (CH_4), nitrous oxide (N_2O) and halocarbons. For several reasons, it is useful to express the contribution of different greenhouse gases in a common metric. First of all, this enables monitoring overall trends in greenhouse gas emissions and comparing the importance of different sources. Secondly, such a metric allows for a determination of possible (economic) trade-offs between reducing different greenhouse gases as part of a multi-gas mitigation strategy. The option to substitute between gases is sometimes referred to as what-flexibility. It has been shown that strategies that allow such flexibility can reach climate objectives more cost-effectively than single-gas mitigation approaches (van Vuuren

© 2015 IOP Publishing Ltd. Environ. Res. Lett. *10 (2) 024001. doi:10.1088/1748-9326/10/2/024001.*
Creative Commons Attribution 3.0 licence (http://creativecommons.org/licenses/by/3.0/). Used with the authors' permission.

et al 2006b, Weyant et al 2006). This was, in fact, already acknowledged by policy-makers in 1997, as the Kyoto Protocol (UNFCCC 1998) was formulated in terms of a multi-gas approach. In addition to the reduction of CO_2, the Kyoto Protocol covers methane, nitrous oxide and a selection of F-gases.

Expressing the contribution of individual gases in one metric is far from straightforward: there are notable differences in radiative properties and atmospheric lifetime between gases. Moreover, many of these properties change over time, as they depend on the composition of the atmosphere. As a result, various metrics have been proposed that all have their strengths and weaknesses in representing the contribution of different gases (see for an overview Fuglestvedt et al 2003). The so-called global warming potential (GWP) is by far the most used metric. However, the GWP is criticized, among others because the value strongly depends on the time span over which the potential is calculated and the inconsistency of the GWP concept with an overall long-term temperature target (Fuglestvedt et al 2000, Smith and Wigley 2000, Manne and Richels 2001, Shine 2009, UNFCCC 2011). The latter may imply that the use of GWPs does not lead to cost-optimal solutions for achieving certain temperature targets (Manne and Richels 2000, O'Neill 2003). Despite the criticism, GWP forms the basis of most multi-gas policies used today, such as the Kyoto Protocol (UNFCCC 1998). The global temperature potential (GTP) (Shine et al 2005) has been proposed as alternative. Proponents of the GTP metric indicate that its link to a temperature target implies that it better relates to the objective of international policies. However, also GTP values depend on particular assumptions in the cause-and-effect chain from emissions to temperature.

Some time ago, the UNFCCC called upon IPCC, and indirectly the research community, to look systematically into the consequences of the use of different metrics and metric values (UNFCCC 2011). The UNFCCC also announced that in the second commitment period of the Kyoto Protocol GWP values of the IPCC AR4 report (IPCC 2007, UNFCCC 2011) will be used, whereas in the first commitment period the GWP values of the IPCC SAR report (IPCC 1995) were used. Several studies have analyzed the impact of different metrics, including GTP and economic based metrics (e.g. Shine et al 2005 Johansson 2012, Reisinger et al 2013). The paper by Reisinger et al

(2013), for instance, discusses the impact of using GTP instead of 100 year AR4 GWPs on global mitigation costs. They found that whereas a fixed 100 year GTP metric would increase costs, time-varying GTPs would reduce costs by about 5% compared to 100 year GWPs. Others have studied metric impacts on costs and emission profiles of multi-gas abatement strategies and find in general small impact on global costs. Smith et al (2013) found, using the GCAM Integrated Assessment Model, that methane emissions vary by at most 18% globally under a range of methane metric weights (4–70) for a fixed carbon price, while global costs increase by 4–23%. Johansson et al (2006) and Aaheim et al (2006) concluded that an optimized emission metric can reduce global costs by several percentage points, compared to an abatement strategy using GWPs. Finally, Godal and Fuglestvedt (2002) found that on a regional scale, the impact on the abatement profile and costs can be significant.

This paper adds to the existing literature by addressing not only the effect of GTP, but also the immediate policy-relevant question of using 100 year GWP values from the different IPCC Assessment Reports and GWPs calculated over different time spans. The impact of different metrics and metric values on both the level and timing of emission reductions of CO_2, CH_4 and N_2O and global abatement costs are analyzed using the FAIR-SiMCaP integrated assessment model, by considering (i) 100 year GWPs from the SAR, TAR and AR4 IPCC reports, (ii) 20 and 500 year GWPs, and (iii) time-varying GTPs. In this way we contribute to the request by UNFCCC (2011) to assess the implications of the choice of metric used to calculate the carbon dioxide equivalent of anthropogenic emissions.

10.2 METHODS

10.2.1 MODELING FRAMEWORK

The FAIR-SiMCaP model (Framework to Assess International Regimes for the differentiation of commitments—Simple Model for Climate Policy Assessment) was used for the analysis (den Elzen et al 2007). This model combines a greenhouse gas abatement cost model with the MAGICC 6 climate model (Meinshausen et al 2011) to calculate long-term emission pathways. FAIR-SiMCaP calculates emission pathways from 2010 to 2100 that achieve climate targets at lowest cumulative discounted abatement costs, using a 5% discount rate (for a sensitivity analysis see the Online Material). The model determines a cost-optimal mix of reduction measures across the emission sources of greenhouse gases covered by the Kyoto Protocol. For this purpose the optimization procedure employs a nonlinear, constrained, optimization algorithm (the MATLAB FMINCON procedure). The optimization procedure optimizes an emission pathway over time, while the substitution metric determines the substitution among gases in any year by multiplying the carbon price with the metric value. The Online Material provides more information on the optimization procedure.

Abatement costs are based on time-dependent and regional information on baseline emissions (see section 2.3) and a set of price-response curves, from now on referred to as marginal abatement cost (MAC) curves. For energy- and industry-related CO_2 emissions, these curves are determined using the TIMER energy model (van Vuuren et al 2007) by imposing a carbon tax and recording the induced reduction in CO_2 emissions. The behavior of the TIMER model is mainly determined by the substitution processes of various technologies based on long-term prices and fuel preferences. These two factors drive multinomial logit models that describe investments in new energy production and consumption capacity. The demand for new capacity is limited by the assumption that capital goods are only replaced at the end of their technical lifetime. The long-term prices that drive the model are determined by resource depletion and technological development. Technological development is determined using learning curves or through exogenous assumptions. Emissions from the energy system are calculated by multiplying energy consumption and production flows by emission factors. A carbon tax can be used to induce a dynamic response, such as an increased use of low- or zero-carbon technologies, energy efficiency improvements and end-of-pipe emission reduction technologies. Negative emissions can be achieved by a combination of the use of bioenergy and carbon capture and storage.

FAIR-SiMCaP captures the time- and pathway dependent dynamics of the underlying TIMER model, that are caused by technology learning and inertia related to capital-turnover rates, by scaling the MAC curves based on the reduction effort in the previous years. The model limits the MAC curves to 1500 \$/tC-eq (409 \$/tCO$_2$-eq), as the underlying TIMER model provides little additional emission reductions above this value.

For non-CO$_2$, the MAC curves of Lucas et al (2007) were used. These are based on MAC curves from the EMF21 project (Weyant et al 2006), but made time-dependent to account for technology change and the removal of implementation barriers, while consistency was ensured by using relative reductions rates compared to a business-as-usual emission level. Moreover, the annual reduction in non-CO$_2$ emissions are assumed to be limited to 2.5%–5% of yearly baseline emissions for most sources, depending on the source (van Vliet et al 2012). These limits are implemented to model the inertia in non-CO$_2$ emission reductions and are based on an estimate of the capital turn-over rate and practices in these sectors. The Online Material provides more information on the shape of the MAC curves and implementation of non-CO$_2$ inertia.

10.2.2 METRIC IMPLEMENTATION

The chosen metric in FAIR-SiMCaP impacts the substitution across the different gases in a single year as it changes the value of the gas vis-à-vis other case. In the model, this is implemented by scaling the MAC curve of each individual gas using the different conversion factors from tons of a specific greenhouse gas to C-equivalent emissions. A change of metric also affects the optimization over time and may result in different optimal emission pathways. For instance, an increase in the methane GWP value from 21 to 25, makes it more attractive to reduce methane. In this example, the price for reaching a certain reduction potential of C-equivalent methane emissions changes by a factor of 21/25. As indicated in the introduction, we have compared the 100 year GWP metric to the time-varying GTP metric and other metrical GWP values.

10.2.2.1 GWP METRIC

In this paper we use GWP values reported by subsequent IPCC reports. GWPs are based on the integrated radiative forcing over a specific time period of a certain greenhouse gas resulting from a 1 kg pulse emission. IPCC follows a methodology assuming an atmospheric background of constant greenhouse gas concentrations. Alternatively, GWP values can be calculated assuming a dynamic atmospheric background concentration. These have the advantage that more of the relevant dynamics are captured, but at the cost of introducing an arbitrary element regarding the development of future emissions.

For determining GWPs, two important inputs are needed: the specific radiative efficiency of a greenhouse gas and its atmospheric lifetime. GWPs are usually expressed relative to the absolute GWP value of CO_2, so the ratio of these two numbers results in a dimensionless GWP value. As there are large differences in the lifetimes of greenhouse gases, GWPs strongly depend on the time span over which the potential is calculated. To cover this, the IPCC Fourth Assessment Report (AR4) quotes 20, 100 and 500 year time spans. The warming potential of CH_4 relative to CO_2 is a factor 9 higher with a 20 year time span than with a 500 year time span (see table 1), due to the short atmospheric lifetime of CH_4.

Current policies mostly use the 100 year GWP values from the Second Assessment Report (IPCC 1995). To explore the potential impact of a change in metric value, we consider the effects of differences in 100 year GWP values between the three IPCC reports SAR, TAR and AR4, but also look at the impact of AR4 values based on a 20 and 500 years time span (IPCC 1995, IPCC 2001, IPCC 2007).

10.2.2.2 GTP METRIC

The GTP is one of the most discussed alternative metrics to the GWP (Fuglestvedt et al 2003, Shine et al 2005). Instead of the integral of the radiative forcing over some fixed time span, the GTP considers the influence of the emission of a greenhouse gas at time t on the global temperature at a certain predefined future year. This effectively makes the GTP time dependent, as the relative impact of different gases varies over time, mostly as result of differences in their atmospheric lifetime. Alternatively, one can also consider a fixed GTP metric of which the associated metric values are very similar to those of a 500 year GWP.

Table 1. GWP values and lifetimes for CH_4 and N_2O used in scenarios, where SAR, TAR and AR4 refer to 100 year GWP values (IPCC 1995, IPCC 2001, IPCC 2007).

	AR4	SAR	TAR	AR4 20 yr	AR4 500 yr	Lifetime (yr)
CO2	1	1	1	1	1	—
CH4	25	21	23	72	7.6	12
N2O	298	310	296	289	153	114

The GTP values used in this paper are based on calculations by Shine et al (2005). The dynamic nature of the GTP values can be easily explained using a simple representation of the climate system (equation (1)).

$$C\frac{d\Delta T}{dt} = \Delta f - \lambda \Delta T \qquad (1)$$

In equation (1), the heat capacity of the climate system is indicated as C (in $J\ m^{-2}\ K^{-1}$), the temperature change as ΔT (in K), the climate sensitivity parameter as λ, which entails climate feedback processes (in $W\ m^{-2}\ K^{-1}$) and the radiative forcing resulting from each of the greenhouse gases as ΔF (in $W\ m^{-2}$). The radiative forcing and temperature change variables are time dependent. Shine et al (2005) solved this equation for ΔT, with a description for the radiative forcing following a pulse emission of each relevant greenhouse gas. This way absolute GTPs are obtained, representing the change in temperature at time t due to an emission of 1 kg of a specific greenhouse gas. As with the GWP, we assume a constant atmospheric background and take the ratio between the GTP of each greenhouse gas to that of CO_2 (see figure 1). Taking this ratio shows the relative temperature effect of the non-CO_2 greenhouse gas with respect to CO_2.

In our analysis of the GTP metric, we only substitute the metric values of CH_4 and N_2O with GTP values, as these gases are the most important in terms of warming and therefore are the main focus of our analysis. For reasons of simplification, we use AR4 100 year GWP values for the other gases HFCs, PFCs and SF6 (note that the GWP scenarios do use the corresponding metric values for all Kyoto gases).

10.2.3 SCENARIOS

As baseline scenario for the analysis, the IMAGE implementation of the OECD Environmental Outlook 2012 is used (OECD 2012). Greenhouse gas emissions in this scenario are driven by factors such as population growth, industrial activity, land-use change and technological development. In the first 30 years, the baseline roughly follows the IEA (2010) baseline scenario; after this period, the baseline follows medium assumptions for population, income and technology development. The scenario results in a rapid increase in greenhouse gas emissions in the next 50 years—followed by a more modest increase later on, reaching a radiative forcing level of 6.7 W m^{-2} in 2100.

The mitigation scenarios are based on a radiative forcing target of 3.5 W m^{-2} in 2100, with a corresponding global temperature change of 2.2 °C relative to pre-industrial, where the same temperature is reached for all metric scenarios, under an equilibrium climate sensitivity of 3 °C. The forcing target covers the contribution of all major greenhouse gases and aerosols as represented in the MAGICC model. This target was chosen as it is a commonly studied policy target, while there is also sufficient flexibility in emission pathways towards this target to assess the importance of different metrics (see also van Vuuren and Riahi 2011). In reaching the long-term target, we allow for an overshoot in the period 2010–2100. To analyze the sensitivity of targets and overshoot on the results, we have performed several sensitivity runs—the results of which are provided in the Online Material.

In our analysis, we focus on CO_2, CH_4 and N_2O, but all Kyoto gases are included in the model calculations. In the mitigation analysis, we use the 100 year GWP values from AR4 as reference scenario and compare those to the scenarios with the alternative metrics.

Because in the FAIR model, mitigation costs in a certain year depend in a nonlinear way on the mitigation pathway, the model can yield different emission pathways that achieve the 3.5 W m^{-2} target at nearly the same cumulative costs. For instance, an early action profile profiting from learning may achieve the same target at similar costs as a profile with some delay profiting from discounting. The nonlinearity and pathway dependencies imply that the optimization routine may report

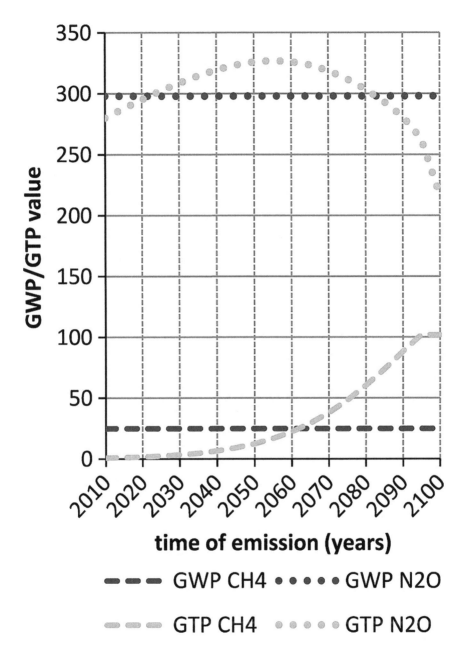

Figure 1. Five year GTP(t) values of CH$_4$ and N$_2$O for a time horizon in the year 2100 based on Shine et al (2005), calculated with IPCC (2007) data, compared to 100 year GWP values from IPCC (2007).

local minima. To account for this, the optimization is run 64 times for every scenario using randomized initial conditions. Some of these runs may yield infeasible solutions due to infeasible starting conditions. We report the results of the set of runs that are within 0.5% of total cumulative discounted costs of the cheapest run (see table 2). In this way, flexibility in the choice of early action versus late response—with similar costs over the whole period—is accounted for. This results in a range of emissions that accounts for the different possible strategies in emission reduction. The emissions range can therefore be interpreted as an indication of flexibility in the most optimal reduction pathway.

10.3 RESULTS

10.3.1 BASELINE

As shown in figure 2, both CH_4 and N_2O emissions increase in the baseline scenario, but only modestly as the growth of main drivers of these emissions (agricultural production, production of fossil fuels) slow down (van Vuuren et al 2006a). The different metric values of GWPs lead to large differences in total C-eq emissions. The time-dependent GTP value leads to a change of relative importance of CH_4 and N_2O with respect to CO_2 and to each other over time. In order to compare emission reductions of different greenhouse gases over time between scenarios, the rest of the paper describes emissions of CH_4 and N_2O in tons of gas.

10.3.2 MITIGATION SCENARIO

10.3.2.1 RESULTS FOR DIFFERENT GWP VALUES (SAR, TAR AND AR4—AND 20, 100 AND 500 YEARS)

Changing between SAR, TAR and AR4 100 year GWP values leads to very small differences in global CO_2, CH_4 and N_2O emissions in the mitigation scenario (see left panels of figure 3). The (small) difference in

Table 2. Number of runs within 0.5% of cumulative discounted costs of the cheapest run for each scenario.

	AR4	SAR	TAR	AR4 20 yr	AR4 500 yr	GTP
#runs	19	18	18	11	22	13

the emissions of methane can be readily understood: since methane has a higher GWP value in AR4 than in SAR and TAR, methane emissions are lowest for AR4 GWP values. As for the impact on emissions, also the impact on global cumulative costs and carbon price are small (see left panel figure 4).

Using 20 year or 500 year GWP values instead of the 100 year GWP values leads to much larger changes in the global CH_4 reduction strategy (right panel of figure 3(b)). As expected, reducing methane becomes more attractive for the 20 year GWP values. For the 500 year GWP values, the opposite is true. Reductions in N_2O emissions exhibit less variation over the different scenarios than methane emissions (right panel of figure 3(c)), as N_2O has a relatively smaller variation in the GWP value than CH_4. The AR4 500 year GWP value for N_2O forms an exception, as it is merely half of the 100 year GWP value (see table 1), resulting in higher N_2O emissions throughout the larger part of the century.

The large difference in impacts on emissions, in particular methane, between 20, 100, and 500 year GWP values not only leads to a different substitution among gases, but also to a change in the overall timing of emission reductions. The early cuts in methane emissions in the 20 year GWP scenario lead to some delay in CO_2 emission reductions of 5–10 years, with emissions about 1 GtC/year higher during the 2035–2060 period. By the end of the century more intensive reductions are required in CO_2 emissions in the 500 year GWP scenario to compensate for the lower reduction of methane. Consequently, CO_2 emissions are clearly lower than in the 100 year GWP scenario.

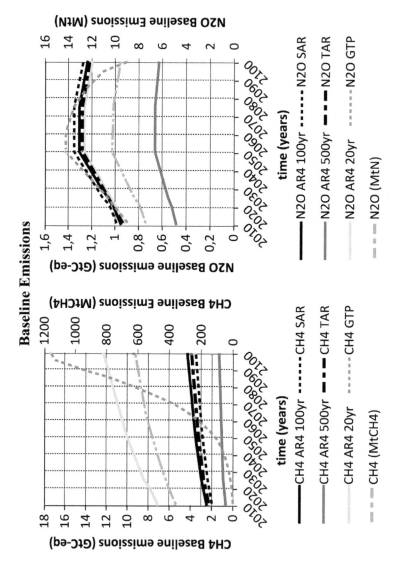

Figure 2. The left panel shows CH$_4$ baseline emissions in GtC-eq on the left vertical axis (for comparison emissions as measured in MtCH$_4$ are shown on the right vertical axis; dashed line only). The right panel shows N$_2$O emissions in GtC-eq on the left vertical axis (again emissions in MtN are shown on the right vertical axis for the dashed line).

Since a large part of the non-CO_2 reductions is relatively cheap, the carbon price dynamics in the first part of the century can be readily understood through the achieved CO_2 reductions: delayed reductions are associated with a delayed increase of the carbon price. The differences in global discounted costs between the scenarios are small, except for the AR4 500 year scenario. This scenario has 20% higher costs than all the other scenarios, while the other five scenarios only differ by a maximum of 6% (see figure 4). In the 500 year GWP scenario, the lower non-CO_2 reductions (or equivalently, the higher CO_2 weight) imply extra CO_2 reductions especially in the second half of the century. These additional reductions drive up the carbon price and therefore increase costs substantially. The carbon price path shows a pathway that reflects, among others, the underlying dynamics in the energy system related to inertia and learning dynamics, model comparison shows that the carbon price pathways across models can vary between simple exponentially increasing pathways, more linear pathways to even stabilizing ones.

10.3.2.2 RESULTS OF USING GTP VALUES

The GTP scenario values methane reductions in particular by the end of the century. The results seem to be a cross-over scenario between the 20 and 500 year GWP scenarios with small methane emission reductions in the period 2010−2050 and a more rapid emission reduction rate for methane in the second half of the century (figure 3(b), right panel). Despite the rapid reduction rate at the end of the century, the CH_4 emission level in 2100 is still higher than in the reference GWP scenario. This is compensated by lower emission levels of N_2O and CO_2 at the end of the century.

Contrary to the result of Reisinger et al and the common assumption that using GTP results in achieving climate targets at lower costs, our GTP scenario results in slightly higher costs than the AR4 100 year GWP scenario. The main reason for this is our assumed restriction on the annual reduction rate of CH_4 emissions by sector. This results in

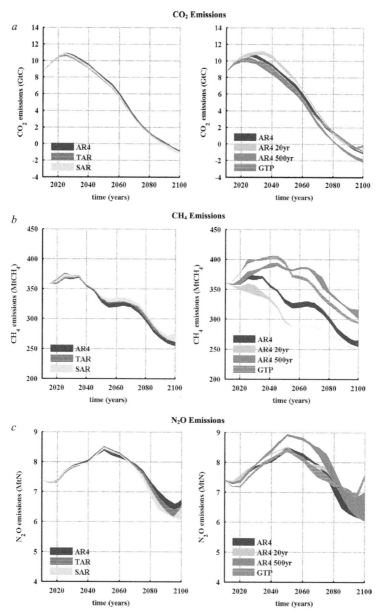

Figure 3. (a)–Emissions of CO_2 (including land-use emissions). (b)—Anthropogenic emissions of CH4. (c)—Anthropogenic emissions of N_2O. Emissions are shown for six scenarios, using 100 year SAR, TAR and AR4 GWP values (left panel) and using 20, 100 and 500 years GWP and GTP values (right panel). Emission range is reported for all scenarios within 0.5% of costs of the cheapest run.

less CH_4 reductions by 2100 with a GTP metric than with the 100 year GWP, even though the CH_4 emissions are valued higher with GTP in equivalent terms at the end of the century. As CH_4 emission reductions are lower using the GTP metric, additional CO_2 reductions are required in this scenario, which (slightly) increases overall costs. Without the restriction on the annual CH_4 reduction rate (compared to the previous year) and other non-CO_2 reductions, the GTP scenario is slightly cheaper than the 100 year GWP scenario, in line with the Reisinger et al result (see Online Material). The FAIR simulation model represents current policies, the abatement of CH_4 and CO_2 is coupled via the CO_2-eq price—and in addition governed by inertia dynamics. The model will not invest in CH_4 abatement before it becomes economic to do so. This is different from a full optimization model that could, depending on the set-up, invest in CH_4 abatement early independent from CO_2 abatement to profit from it later in the century.

All scenarios reach a temperature of 2.2 °C by 2100 under a radiative forcing target of 3.5 W m^{-2}. The transient temperature and forcing levels over the century are very similar, with a maximum difference of 0.1 °C and 0.19 W m^{-2} in all years for all different metrics.

As a sensitivity analysis, we performed the 100 year GWP and GTP scenarios under different climate targets (both forcing and temperature targets), which are presented in more detail in the Online Material. The scenario under a temperature target of 2.2 °C with a GTP emission metric leads to very similar costs and emissions as the same scenario under a radiative forcing target of 3.5 W m^{-2}. The 100 year AR4 GWP and GTP scenarios exhibit similar behavior for different climate targets. For a forcing target of 2.8 W m^{-2}, a 3.5 W m^{-2} target without the option of an overshoot and a temperature target of 2 °C, the GTP scenario shows higher costs, while the emission dynamics are similar to a 3.5 W m^{-2} scenario, albeit with different emission levels. A more ambitious climate target (i.e. a 2 °C and a 2.8 W m^{-2} target) leads to smaller differences in CO_2 emissions between the two scenarios compared to a 3.5 W m^{-2} target, whereas the inability to overshoot the climate target during the century induces earlier reductions. Also, scenarios under discount rates of 3% and 7% show similar dynamics to the scenario under a discount rate of 5%, although emission and cost differences are

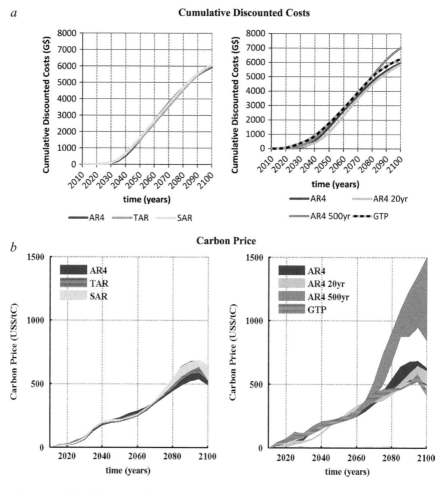

Figure 4. (a)—Climate policy cumulative discounted costs for the cheapest run of each of the six scenarios, using 100 year SAR, TAR and AR4 GWP values (left panel) and using 20, 100 and 500 years GWP and GTP values (right panel). The discount rate was set to a value of 5%. (b)–Carbon price for six scenarios, using 100 year SAR, TAR and AR4 GWP values (left panel) and using 20, 100 and 500 years GWP and GTP values (right panel). The carbon price has units of 2005 US$ per ton C-equivalent. Price range is reported for all scenarios within 0.5% of costs of the cheapest run.

smaller between 100 year AR4 GWPs and GTPs for the lower discount rate (see Online Material).

10.4 CONCLUSIONS AND DISCUSSION

10.4.1 THE CHOICE OF THE METRIC MOSTLY IMPACTS THE TIME PROFILE OF METHANE REDUCTIONS

As different GWP values assign a different value to CH_4 emissions, the change of metric also changes the emission reductions of this gas. Obviously, these changes become relevant if the differences between the metric values are substantial enough. In our study, this is the case for the AR4 20, 100 and 500 year GWPs and the GTP scenario. These changes in CH_4 emissions also affect the timing and depth of CO_2 reductions. However, indirect reductions in CH_4 due to efforts to mitigate CO_2 emissions, for instance a decrease in coal use, are not included in the calculations. This type of interaction can lessen the impact of a change of metric value, as discussed in Smith et al (2013).

Nitrous oxide emission reductions show less variation than methane reductions when applying different GHG weights, as a result of the smaller relative differences among N_2O metric values compared to those for CH_4.

The radiative forcing target as chosen in this study (3.5 W m^{-2}) allows for some flexibility in the timing of emission reductions; for more stringent climate targets this flexibility can be less (see Online Material).

10.4.2 THE GLOBAL MITIGATION PROFILE AND ASSOCIATED GLOBAL COSTS ARE FOUND TO BE NOT VERY SENSITIVE TO THE CHANGES IN METRIC VALUES OF CH_4 AND N_2O AS REPORTED IN SAR, TAR AND AR4

The most important reason for this is that the differences between the metric values reported in the subsequent IPCC reports are relatively small: the largest difference is about 20% for the GWP of methane. The Fifth Assessment Report states an increase in GWP metric values for CH_4 and a decrease for N_2O (IPCC 2013). This will in principle lead to somewhat higher CH_4 and lower N_2O reductions. The changes in metric values are expected to have a minor impact on emission reductions and costs, although the change in N_2O GWP values is somewhat larger than between previous assessment reports.

Additionally, the CH_4 abatement cost curve as used in our model plays a role: as a considerable amount of the reductions is relatively cheap, differences in CH_4 mitigation only slightly affects total abatement costs. Moreover, at the high end of the curve a considerable part of the methane emissions cannot be reduced (as a result of lack of abatement potential); here also the abatement decision is independent of the metric choice. The small differences in emissions among these scenarios result in minor global cost differences.

10.4.3 USING GWP VALUES CALCULATED OVER DIFFERENT TIME SPANS HAS A MODERATE IMPACT ON GLOBAL COSTS

The difference in impact of using 20 and 100 year GWP values on global costs is relatively small (up to 4%). However, the 500 year GWP values lead to significantly higher costs (of 18%), as non-CO_2 GHGs are given such low weights that considerable additional CO_2 reductions are necessary. In other words, the importance of CO_2 for reaching a radiative forcing target at the end of the century is so large that the exploitation of the CH4 mitigation potential is much lower than in the other GWP scenarios, leading to higher costs. Reisinger et al (2013) also find higher (up to 10%) costs for their fixed GTP scenario, which uses a CH4 weight comparable to a 500 year GWP weight.

10.4.4 THE IMPACT OF DIFFERENT METRIC VALUES ON TEMPERATURE IS VERY SMALL

All scenarios yield the same temperature change relative to pre-industrial of 2.2 °C in 2100. The differences in temperature change and radiative forcing during the century among the scenarios are small (up to a maximum temperature difference of 0.1 °C).

10.4.5 IMPLEMENTING A TIME-VARYING GTP METRIC DOES NOT NECESSARILY LEAD TO LOWER COSTS, DEPENDING ON INERTIA AND OTHER NONLINEAR IMPACTS ON THE METHANE REDUCTION RATE

Using a GTP metric leads to higher CH_4 emissions at the end of the century compared to the 100 year GWP scenario, due to limitations to the annual reduction of non-CO_2 emissions implemented in the FAIR model, even though in the GTP scenario methane has a higher associated weight at that time. Without these limitations on annual reductions, GTP does lead to lower costs than 100 year GWPs, confirming the results by for instance Reisinger et al (2013). Although it is uncertain how large the inertia in reducing non-CO_2 emissions is, the addition of inertia in reducing non-CO_2 emissions leads to higher costs in general, and specifically when using the GTP metric. So the advantage, or disadvantage, of a GTP metric depends on the speed by which methane can be reduced. Future work on the speed by which non-CO_2 emissions can be reduced is therefore warranted.

10.4.6 IN THIS PAPER WE FOCUS ON GLOBAL RESULTS ONLY. REGIONAL IMPACT OF DIFFERENT METRICS OR METRIC VALUES MIGHT DIVERGE FROM GLOBAL RESULTS, DEPENDING ON THE RELATIVE CONTRIBUTION OF NON-CO_2 GHGS TO TOTAL EMISSIONS AND EMISSION TRADING

Some regions have particular large shares of CH_4 emissions. For these regions, more substantial impacts for regional costs can be expected. The impacts are likely to be strongly dependent on the climate policy regime: different metric values will not only change relative abatement costs, but also the allocation of emission permits and resulting emission trading.

REFERENCES

1. Aaheim A et al 2006 Costs savings of a flexible multi-gas climate policy Energy J. 27 485–501
2. den Elzen M G J et al 2007 Multi-gas emission envelopes to meet greenhouse gas concentration targets: costs versus certainty of limiting temperature increase Glob. Environ. Change 17 260–80
3. Fuglestvedt J S et al 2000 Climate implications of GWP-based reductions in greenhouse gas emissions Geophys. Res. Lett. 27 409–12
4. Fuglestvedt J S et al 2003 Assessing metrics of climate change—current methods and future possibilities Clim. change 58 267–331
5. Godal O and Fuglestvedt J 2002 Testing 100 year global warming potentials: impacts on compliance costs and abatement profile Clim. Change 52 93–127
6. IEA 2010 World Energy Outlook 2010 (Paris, France: International Energy Agency)
7. IPCC 1995 Climate Change 1995, IPCC Second Assessment (Cambridge, UK: Cambridge University Press)
8. IPCC 2001 Climate Change 2001.The Science Of Climate Change (Cambridge, UK: Cambridge University Press)
9. IPCC 2007 Climate Change 2007: The Physical Science Basis. Contribution of Working Group I to the Fourth Assessment Report of the Intergovernmental Panel on Climate Change (Cambridge and New York: Cambridge University Press)
10. IPCC 2013 Climate Change 2013: The Physical Science Basis. Contribution of Working Group I to the Fifth Assessment Report of the Intergovernmental Panel on Climate Change (Cambridge and New York: Cambridge University Press)
11. Johansson D J A 2012 Economics- and physical-based metrics for comparing greenhouse gases Clim. Change 110 123–41
12. Johansson D J A et al 2006 The cost of using global warming potentials: analysing the trade off between CO2, CH4 and N2O Clim. Change 77 291–309
13. Lucas P et al 2007 Long-term reduction potential of non-CO2 greenhouse gases Environ. Sci. Policy 10 85–103
14. Manne A S and Richels R G 2000 The Global Carbon Cycle: Integrating Humans, Climate, and the Natural Word ed C B Field and M R Raupach (Washington DC: Island Press) A multi-gas approach to climate policy. With and without GWPs
15. Manne A S and Richels R G 2001 An alternative approach to establishing trade-offs among greenhouse gases Nature 410 675–7
16. Meinshausen M et al 2011 Emulating coupled atmosphere-ocean and carbon cycle models with a simpler model, MAGICC6: I. Model description and calibration Atmos.Chem. Phys. 11 1417–56
17. O'Neill B C 2003 Economics, natural science and the costs of global warming potentials Clim. Change 58 251–60
18. OECD 2012 OECD Environmental Outlook to 2050: The Consequences of Inaction (Paris: Organisation for Economic Co-operation and Development)
19. Reisinger A et al 2013 Implications of alternative metrics for global mitigation costs and greenhouse gas emissions from agriculture Clim. Change 117 1–14

20. Shine K P 2009 The global warming potential-the need for an interdisciplinary retrial Clim. Change 96 467–72

21. Shine K P et al 2005 Alternatives to the global warming potential for comparing climate impacts of emissions of greenhouse gases Clim. Change 68 281–302

22. Smith S J et al 2013 Sensitivity of multi-gas climate policy to emission metrics Clim. Change 117 663–75

23. Smith S J and Wigley T M L 2000 Global warming potentials: I. Climatic implications of emission reductions Clim. Change 44 445–57

24. UNFCCC 1998 Kyoto Protocol to the United Nations Framework Convention on Climate Change. (Bonn: United Nations Framework Convention on Climate Change)

25. UNFCCC 2011 Greenhouse Gases, Sectors And Source Categories, Common Metrics To Calculate The Carbon Dioxide Equivalence Of Anthropogenic Emissions By Sources And Removals By Sinks, And Other Methodological Issues (Durban: United Nations Framework Convention on Climate Change)

26. van Vliet J et al 2012 Copenhagen accord pledges imply higher costs for staying below 2 °C warming: a letter: a letter Clim. Change 113 551–61

27. van Vuuren D and Riahi K 2011 The relationship between short-term emissions and long-term concentration targets Clim. Change 104 793–801

28. van Vuuren D P et al 2006a Long-term multi-gas scenarios to stabilise radiative forcing—exploring costs and benefits within an integrated assessment framework. energy journal, multi-greenhouse gas mitigation and Energy J. 27 201–34

29. van Vuuren D P et al 2006b Multigas scenarios to stabilise radiative forcing Energy Econ. 28 102–20

30. van Vuuren D P et al 2007 Stabilizing greenhouse gas concentrations at low levels: an assessment of reduction strategies and costs Clim. Change 81 119–59

31. Weyant J P et al 2006 Overview of EMF—21: multi-gas mitigation and climate change Energy J. 27 1–32 multi-greenhouse gas mitigation and climate policy

PART V

CONCLUSIONS

CHAPTER 11

A New Scenario Framework for Climate Change Research: The Concept of Shared Climate Policy Assumptions

ELMAR KRIEGLER, JAE EDMONDS, STÉPHANE HALLEGATTE, KRISTIE L. EBI, TOM KRAM, KEYWAN RIAHI, HARALD WINKLER, AND DETLEF P. VAN VUUREN

11.1 INTRODUCTION

Scenarios about future socioeconomic and climate developments are used to study the scope and implications of climate change and responses to it (Nakicenovic et al. 2000; Moss et al. 2010; Kriegler et al. 2012; van Vuuren et al. 2012). Over the past years, the research community has pursued a vision for an improved scenario-based climate change assessment process (Moss et al. 2008). An important feature of that vision was the ability to explore a greater variety of socioeconomic development pathways for emissions mitigation and climate impacts and adaptation. To that end Representative Concentration Pathways (RCPs) were created (van Vuuren et al. 2011) as a preliminary step to facilitate the creation of a set of climate model calculations, which in principle could be matched with new scenarios sharing the same climate forcing. To link possible

© *The Author(s) 2014. Licesnsed by SpringerLink.* Climatic Change, *February 2014, Volume 122, Issue 3, pp 401-414. doi: 10.1007/s10584-013-0971-5. Creative Commons Attribution License.*

socioeconomic development futures with a number of different climate outcomes, assumptions about mitigation and adaptation climate policies are needed. In this way, the analysis of climate policy is deeply embedded in a scenario-based assessment of climate change.

It is a fundamental question how scenarios can best be used to analyze policy (Morgan and Keith 2008). In general, this will depend on the policy context. If the decision problem can be broken down to the options and contingencies a specific decision maker is confronted with, scenarios may be developed for each alternative course of action and conceivable state of the world in close interaction with the decision makers (e.g. Parson 2008). Examples include for instance scenarios developed to inform specific business or planning decisions. Climate policy analysis is different in that it relates to a wider set of questions, ranging from regional adaptation plans for the near term to short- and long term climate mitigation strategies as, e.g., considered in international climate policy negotiations. Climate change scenarios are therefore often targeted to intermediate users, such as researchers who use scenarios as inputs into their work (Moss et al. 2008). These scenarios have to be specified in a way that retains their flexibility to be applied to different climate policy contexts. There is sometimes a perceived tension between scenarios directed to research vs. decision support. However, they are rather complementary in nature. Scenarios directed to research should be applicable as analytic tools to develop decision support scenarios in a specific context.

This paper takes up the question of scenario based policy analysis in the context of the new scenario framework. The framework is introduced in a series of four papers in this special issue (Ebi et al. 2014; van Vuuren et al. 2014; O'Neill et al. 2014, are the other three). The general approach is presented in van Vuuren et al. (2014). At its core is the concept of a scenario matrix that combines so-called shared socioeconomic reference pathways (SSPs) with climate forcing outcomes as described by the representative concentration pathways (RCPs, van Vuuren et al. 2011). The RCPs reach different levels of climate forcing in the year 2100 and thus can serve as proxy for climate targets.[1] The SSPs are introduced in O'Neill et al. (2014) and aim to characterize socioeconomic challenges to mitigation and adaptation in a reference case without explicit climate policies and without consideration of climate change impacts. Per definition, the SSPs provide the constituents of a reference scenario that can serve as a counterfactual to evaluate the impact of climate policy. To allow their broad applicability they have to exclude any climate policy, but can include other policies that are not directly related to climate. In fact, to be useful for climate policy analysis they should include all those policies controlled by non-climate objectives that will either have a substantial impact on climate policy related outcomes or be substantially impacted by climate policy itself.

The cells of the scenario matrix are defined by combinations of distinct SSPs and climate forcing outcomes (as characterized by the RCPs). Some of these cells at the higher end of the forcing range, e.g. for the two RCPs attaining 6 and 8.5 W/m^2 by the end of the century, may be populated by reference scenarios that project socioeconomic developments and emissions in the absence of climate policy and are based purely on the SSPs, although the extent and pattern of climate change could affect development pathways such that the SSPs may need to be modified when creating a scenario. Climate policies will be needed to an increasing degree to reach the lower part of the RCP range, particularly for 2.6 and 4.5 W/m^2. The scenarios created in these matrix cells will therefore draw on SSPs, but also make assumptions about climate policy including mitigation and adaptation measures. The impact of such policy scenarios could then be analysed by comparing it with the reference scenario for a given SSP (within a column of the matrix; see van Vuuren et al. 2014, for further discussion).

This is the entry point of our analysis. We ask the question how the formulation of climate policy scenarios could be framed to enhance their applicability to the scenario framework, in particular concerning their introduction into the RCP-SSP matrix. We note that our discussion applies entirely to the use of climate policy scenarios in this framework, not to climate policy scenarios in general. There always has been, and always should be a large variety of climate policy scenario analyses to cater to the widely different decision contexts in the area of climate change.

We start from two observations. Firstly, the RCPs, as a measure of the anthropogenic forcing of the climate system, may characterize the target of mitigation action, but not the type and structure of climate policy interventions including mitigation and adaptation measures. Secondly, any set of climate policy assumptions will have strong implications for the outcome of the scenario analysis in the proposed RCP-SSP framework. For example, the assumption of the time by when a long term climate target is adopted globally will strongly influence the ability to reach this target in addition to the challenges to mitigation characterized in the SSPs. Likewise, the existence of a global adaptation fund and international insurance mechanisms against climate change impacts will affect the ability to implement adaptation measures in addition to the challenges to adaption in the SSPs. We therefore ask the question how the SSP and forcing (RCP) dimensions should be augmented by Shared Climate Policy Assumptions (SPAs) to better incorporate the climate policy dimension within the scenario framework. The idea that policy assumptions may be shared between studies should not be confused with a prescription that all countries should take the same level of action, or should act under internationally co-ordinated regimes. To this end, we note that assumptions are not recommendations, and that the usefulness of making assumptions will depend on how broad a range of plausible climate policy formulations is assessed. This may include, at a minimum, climate policy assumptions that describe regionally differentiated and globally harmonized approaches.

Section 2 will discuss the definition and scope of SPAs, while Section 3 conceptualizes their quantitative and qualitative elements. Section 4 turns to the difficult question of the dividing line and the interdependence between SPAs and SSPs. Section 5 provides examples of how SPAs can be integrated as a third dimension in the scenario matrix architecture. Section 6 concludes that it can be useful to characterize the key dimensions of climate policy assumptions in SPAs much in the same way as socioeconomic reference assumptions are summarized in the SSPs.

11.2 DEFINING SHARED CLIMATE POLICY ASSUMPTIONS

Climate policies can be characterized in terms of their attributes, such as the stringency of policy targets, the set of climate policy instruments employed to achieve the targets , and the time and place in which climate policy instruments are deployed. Following a proposal by Kriegler et al. (2012), we define Shared Climate Policy Assumptions as capturing key characteristics of mitigation and adaptation policies up to the global and century scale. The latter requirement emerges from the fact that SPAs should relate to the global scenario framework. At the same time they should be flexible enough to allow for regional differentiation of policies.

Concretely, SPAs should describe three attributes of climate policies (Kriegler et al. 2012). The first attribute is the global (collection of) "climate policy goals" such as emissions reductions targets, or different levels of ambition in limiting residual climate damages, e.g. in terms of development indicators that should not be jeopardized by climate change. However, there is a clear overlap of mitigation policy goals with the forcing (RCP) dimension of the scenario matrix, and of adaptation policy goals linked to achieving sustainable development and other societal goals described in the SSPs. RCP forcing levels can be used to directly describe long term mitigation targets (although not how to achieve those targets), and in any case have strong implications for the admissible global outcome of regional emissions reduction targets.

Thus, we have to distinguish two types of SPAs: a full SPA that includes all mitigation and adaptation policy targets, and thus embeds the forcing (RCP) dimension and possibly aspects of an SSP in it. And a reduced SPA that excludes the mitigation policy goals, at least as far as they relate to emissions reductions and global concentration and forcing outcomes, and the adaptation policy goals as far as they relate to development goals, and therefore is orthogonal to both the forcing (RCP) and the SSP axis of the matrix framework. Thus, the reduced SPA has to be used if variations of policy assumptions for a given RCP-SSP combination are to be explored. In this way, it adds a third axis to the scenario matrix. Both concepts— reduced and full SPA—can be entertained simultaneously. A full SPA may simply be the combination of a reduced SPA and an RCP forcing level. Or, if the mitigation targets are to be specified in terms of regional emissions reductions commitments, the RCP forcing level would determine the global cumulative amount of permitted emissions (and possibly some properties of the global emissions pathway), while the reduced SPA is free to include any distribution of these emissions permits across regions that adds up to the global total. Similarly, for adaptation policy goals, a full SPA would embed SSPs elements, such as development goals that are relevant for adaptation policies, while a reduced SPA would focus purely on the adaptation policy aspect. We note that this differentiation of full and reduced SPAs to ensure orthogonality of the SPA, SSP, and forcing (RCP) axes goes beyond what was proposed in Kriegler et al. (2012).

Second, the shared climate policy assumption should describe the characteristics of the global (collection of) " policy regimes and measures" introduced to reach the policy goals- On the mitigation side, such policy measures could be globally harmonized or regionally differentiated carbon taxes, an international emissions trading scheme with a particular burden sharing mechanism, a mix of different policy instruments ranging from emissions pricing to low carbon technology subsidies to regulatory policies, or a mix of different approaches in different sectors, e.g. including transport policies and schemes to protect tropical forest. SPAs may also relate to mitigation policy dimensions that are often not considered by model studies, but elsewhere in the literature. An example is how particular policies would be financed in practice (Clapp et al. 2012; Winkler et al. 2009). On the adaptation side, the SPA package may include, for example, the type of adaptation measures that are implemented (e.g. more efficient irrigation techniques or water recycling technologies) and the availability of various amounts of international support for adaptation in developing countries.

Third, a shared climate policy assumption should include the "implementation limits and obstacles" to the extent they are considered and are not part of an SSP. Those obstacles could be specified in terms of the exclusion of several policy options for some regions and sectors where they do not appear to be feasible. Such conditions could evolve over time such that initial limits might dissipate or intensify. For example, several land pools may be excluded from carbon pricing due to practical constraints of implementing such a pricing policy. Or a group of regions may be assumed to remain outside an international climate policy regime until some point in time. Likewise, adaptation effectiveness may differ in a model where behavioral biases in risk perceptions are accounted for, or where political economy or enforcement constraints makes it impossible to implement some policies (e.g., land-use regulations aiming at reducing disaster losses are difficult to implement in areas without official land tenure such as informal settlements).

Care needs to be taken to separate implementation limits and obstacles that are attributes of a SPA, and climate policy obstacles that are inherent in the socioeconomic reference environment described in an SSP. The latter can include market distortions, e.g. in energy and labor markets. Babiker and Eckaus (2007) and Guivarch et al. (2011) show that taking into account unemployment and friction in labor market adjustments can change in a significant way the assessment of mitigation costs. In developing countries in particular, existing economic distortions cannot be disregarded in the design of climate policies. The high level of unemployment (often over 25 %), the large share of the informal economy, and the difficulty to enforce regulation (e.g., in terms of land-use planning) and raise taxes create specific difficulties for climate policy. To the extent that such elements are considered, they will be part of an SSP rather than a reduced SPA. However, they could be added to a full SPA that aims to provide a complete picture on the three attributes of climate policy, i.e. goals, instruments, and obstacles.

The formulation of SPAs on a global and century scale can become complex. However, the goal of SPAs should not be to describe the climate policy landscape in every conceivable detail, but rather to summarize and make explicit the central policy assumptions that have to be made anyway by individual studies to produce climate policy scenarios. Seen through this lens, SPAs have been used for quite some time, for example, in integrated assessment model inter-comparison studies (e.g. Clarke et al. 2009; Riahi et al. 2014; Kriegler et al. 2014a) that needed to harmonize their climate policy assumptions in order to make results comparable. It is more and more recognized that such harmonization does not only involve the specification of a long term climate target as would be captured by the RCPs, but also, e.g., the basket of greenhouse gases and sectors to become subject to emissions pricing, the degree of global cooperation and the degree of overshoot of long term targets (Blanford et al. 2014).

A controlled variation of key climate policy assumptions, but not of all the details, will be required to make the scenario matrix approach fully operational. Knowledge about what type of SPAs have been used in modeling will be required to make appropriate comparisons across model analyses or other studies that explore a given combination of SSP and radiative forcing target (RCP). For instance, mitigation cost assessments by two models cannot be compared directly, if these two models make different assumptions on practical obstacles to policy implementation (e.g., if one model assumes a globally integrated carbon market and the other a regionally fragmented mitigation regime). Furthermore, the choice of SPAs will interact with the socioeconomic challenges to mitigation and adaptation that are engrained in the SSPs (O'Neill et al. 2014). For example, whether or not global coverage of emissions reductions can be achieved will affect the ability to reach a prescribed RCP forcing level. Likewise, whether or not a global adaptation fund is put in place will affect the ability of individual regions to adapt to climate change.

11.3 ELEMENTS OF SHARED CLIMATE POLICY ASSUMPTIONS

Shared Climate Policy Assumptions can contain qualitative and quantitative information. The qualitative information consists of a narrative that describes the world of climate policies and their evolution over time and across space (Hallegatte et al. 2011). A key characteristic of mitigation policies is the number of countries that participate in an international climate policy regime over time and the stringency of their commitments and actions. IAM studies often distinguish a benchmark case of fully cooperative action starting immediately with more plausible policy scenarios that include regionally and sectorally fragmented climate policies, staged accession to a global climate regime, and non-participation (Clarke et al. 2009; Blanford et al. 2014; Kriegler et al. 2014b; Luderer et al. 2014). Thus, the SPA narrative should include information on the different timing of participation of regions and nations in emissions mitigation regimes, as well as being explicit whether mitigation stringency is globally uniform or differentiated across regions and countries. It could also contain information about the nature of climate policies—e.g., preferences for fiscal as opposed to regulatory policies, differences in the nature of policies to mitigate fossil fuel and land-use change emissions, and emphasis on behavioral changes, efficiency and demand-side measures vs focusing mitigation more on upstream technology solutions for energy supply. In addition, SPA narratives could take into account information about the constraints and obstacles for mitigation policy.

Key qualitative information on adaptation policies includes institutional policies that are implemented to support adaptation, such as the implementation of a technology transfer agreement at the international scale; the quality of adaptation governance processes (e.g., corruption, capture by interest groups); and the effectiveness of policy implementation (e.g., enforcement of building norms and land-use regulations). Table 1 presents an illustrative example of key components of such narratives, and how they can be combined to a limited number of SPAs that cover a fairly wide range of different climate policy futures.

SPAs will in general also contain quantitative information. As far as full SPAs are concerned, the long term mitigation target as determined by the long term forcing in an RCP will be a central part of this information. In principle, the target could be specified in a number of ways, ranging from a global temperature target, to a climate forcing target to a cumulative emissions budget for the entire world. It may also include some constraints on the pathway of climate forcing or global emissions. The RCPs will be a determining factor for globally aggregate descriptors of mitigation effort even if no global mitigation target is assumed. For example, they will strongly influence (together with the choice of SSP) the global level of a carbon tax path that could be assumed instead of a target (Calvin et al. 2012). Quantitative information that would be part of a reduced SPA, and thus would need to be complementary to information constrained by the RCPs can include, e.g., the allocation of emissions permits to different regions in terms of shares of global emissions (Tavoni et al. 2014), carbon price differentials between regions, sectors and land pools, a timetable for staged accession to a global climate policy regime (Clarke et al. 2009), regional low carbon technology targets (Kriegler et al. 2014b), and land use related policies such as forest protection and bioenergy constraints (Calvin et al. 2014). Quantitative assumptions on adaptation policy can include, for example, adaptation targets such as protection against 100 year flood or drought events, timetables for implementing regional adaptation plans, and the size of an international adaptation fund that is set up to assist countries that are most affected by climate change.

In summary, the SPAs should contain information that is instructive to both the integrated assessment modeler trying to develop a climate mitigation scenario, and an IAV researcher trying to analyse the

Table 1. Illustrative example of a set of reduced SPAs (columns) defined by a collection of policy attributes (rows) SPAs. Those SPAs can be combined with an RCP forcing level and an SSP, although not all combinations can be expected to be consistent and yield a feasible scenario. In addition, it may be useful to consider a "no new policy" SPA that only contains existing climate policies until their time of expiration, assuming no new climate policies thereafter.

Policy attribute	Reference policy	Cooperation & moderate adaptation	Middle road & aggressive adaptation	Fragmentation & moderate adaptation
Mitigation: Level of global cooperation (e.g. measured in terms of average share of emissions under a global climate target over some period)	Low	High	Medium	Low
Mitigation: Start of global cooperation (e.g. the time the first group of countries adopts a global target or an international carbon tax)	Never	Early	Mid Term	Late
Mitigation: Sectoral coverage	Focus on electricity and industry sectors. No significant inclusion of land use based mitigation options.	Carbon pricing on land. Full coverage of energy supply and end use sectors.	Forest protection and bioenergy constraints. Energy supply, transport and industry covered.	Limited forest protection, no limitation on bioenergy use. Electricity and industry covered.
Adaptation: Capacity building (e.g. measured in terms of the size of a global adaptation fund)	Small	Moderate	Large	Moderate
Adaptation: International insurance (e.g. measured in terms of the amount of climate impact insurance available between countries)	Only via international markets, with limited access for some countries	Insurance available for least developed countries	Global insurance provided	Only via international markets, with limited access for many countries

vulnerability to climate change, the costs and benefit of adaptation measures and the residual climate impacts. The level of detail to which it is useful to specify information in the SPAs will depend on the application. Obviously, SPAs at the global and century scale have to be very generic by construction, since a detailed formulation of the global climate policy landscape in 2050 would carry little meaning. As for the case of the SSPs, there is an inherent tension between establishing comparability of different studies and comprehensive coverage of plausible SPAs (compare O'Neill et al. 2014). It may be useful to distinguish between basic SPAs that only include high level information on the scope of mitigation and adaptation actions and thus can summarize a larger set of climate policy studies, and extended SPAs that allow to better control climate policy assumptions as for example adopted in IAM comparison projects. Furthermore, the connection between global studies and mitigation and adaptation analysis conducted at the national level needs further attention. Global SPAs should be flexible enough to be adapted and changed when applied to more short-term and more local/national analysis, much in the same way as global SSPs should allow their adaptation to studies on a local/national level. National SPAs could, for example, be taken up in SPA extensions. However, a set of basic SPAs on the global level will be most useful if it is limited in number, generic in character and broad enough to allow a comprehensive exploration of the climate change scenario space.

11.4 THE RELATIONSHIP BETWEEN SPAS AND SSPS

As discussed above, climate policy assumptions will not be included in SSPs by their definition in terms of socioeconomic reference assumptions. Thus, assumptions about climate policies, even if they correspond to currently planned legislation, should always be part of an SPA rather than an SSP. The dividing line between assumptions on climate policies to be included in SPAs and broader development policies to be included in SSPs will be difficult to draw in many cases. In general it is the motivation for the policy intervention and not the policy itself that determines whether the policy is a climate policy or a policy directed toward another end. As a consequence a policy may be included as a non-climate policy in the SSP, but tightening of the policy in support of greenhouse emissions mitigation, for example, could be included in the SPA. A renewable energy portfolio standard, for example could occur in a reference scenario in support of improved energy security, and a more stringent implementation might be included in an SPA. In the end the test is, would the policy and its stringency be expected to be enacted in the reference, no-climate-policy scenario? If the answer is yes, it belongs in the SSP. To the extent that the policy is deployed and/or tightened only in the mitigation scenario, it belongs in the SPA.

For example, any policy that directly constrains or taxes the emissions of greenhouse gases falls into a SPA. The same holds for any policy that directly addresses adaptation to climate change such as the implementation of an international adaptation fund. In contrast, most development policies such as improving energy access, urban planning, infrastructure, health services, and education are motivated in their own right, and thus are not climate policies. Those policies are part of the socioeconomic reference scenario, and their outline should be included in the SSPs. Such policies may, of course, affect climate policies, or be affected by them, which reinforces the case for their inclusion in SSPs. Care must be taken when combining SPAs with SSPs to ensure consistency of the full policy package (see below). For example, development policy assumptions in the SSPs may have to be adjusted when being combined with an SPA. However, this also holds true for other variables in the SSPs, such as land and energy use patterns that will be affected by climate policy.

There are also borderline cases due to the fact that policies are often motivated by multiple objectives (Linnér et al. 2012; Winkler et al. 2008). Is a renewable portfolio standard motivated by concerns about climate change or energy security or both? Does increased disaster preparedness stem from a concern about the increased frequency or magnitude of such disasters in a changing climate, or does it stem from the objective to decrease the vulnerability of the society to present climate variability? Such cases cannot fully be decided, and one may have to resort to ad hoc judgments on a case-by-case basis. The main point is that the relevant policy assumptions underlying the socioeconomic reference and climate policy scenarios are clearly allocated to either an SSP or a SPA. A clear separation of policies with respect to the climate and non-climate objectives in the SSPs and SPAs will make the framework also useful for the assessment of potential synergies and tradeoff between climate and other non-climate policies (McCollum et al. 2011).

Another important question is how to deal with climate policies and measures that are already implemented and affect the socioeconomic development on a larger scale. The price on greenhouse gas emissions in Europe, imposed directly via the European Emissions Trading System (EU ETS), or implicitly via sectoral measures aiming to reach the targets under the Kyoto Protocol, are a case in point. If such implemented climate policy measures are excluded from the socioeconomic reference scenario, it would have already diverged from reality. In order to avoid this, existing climate policies may be collected in a minimal "existing policy" SPA that can be combined with an SSP when developing the reference scenario. Seen in this way, any socioeconomic scenario, including the reference case, would be based on some SSP and SPA. Only a counterfactual no climate policy scenario would not require the adoption of a SPA, or, put differently, adopt an empty SPA.

The idea of a minimal "existing policy" SPA immediately raises the question of what is an existing climate policy: measures in effect like the EU ETS; or a policy foreseeing future measures that is coded into law like the EU Climate and Energy Package? And how should such policies be projected into the future in a policy reference case defined as continuation of current levels of ambition (Blanford et al. 2014; Kriegler et al. 2014b, Luderer et al. 2014)? These questions will need to be further addressed during the construction and testing of SSPs and SPAs, and their use in the development of socioeconomic scenarios. They are closely related to the long-standing discussion in many fields and applications of what should count as the baseline or reference case, against which actual policy proposals are measured. SPAs provide the flexibility to include different interpretations of "baseline" in the analysis.

When combining SSPs and SPAs to derive a socioeconomic climate policy scenario, care needs to be taken that their combination is consistent. First, SSPs will contain reference assumptions that are affected by climate policies, and those would need to be adjusted to take into account the information in the SPA. Second, some reference assumptions in an SSP, e.g. development policies, will have implications for climate policy, and consistency between a SSP and a SPA would need to be ensured. For example, a narrative describing a regionalized development in a fragmented world can hardly be paired with the assumption of a global carbon market. Likewise, implementation obstacles, e.g. in terms of land pools that can or cannot become subject to carbon pricing, can be more or less consistent with different SSP. It therefore will be the case that not all SPAs can be combined consistently with all SSPs.[2] It is important that at least one SPA be developed so as to be consistent with each SSP.

11.5 INTEGRATING SPAS INTO THE SCENARIO MATRIX ARCHITECTURE

The explicit introduction of a climate policy dimension as captured in the SPAs offers the flexibility to explore adaptation and mitigation policies for different combinations of SSP, RCP and reduced SPAs. For example, Fig. 1 shows the combination of RCPs with reduced SPAs describing different types of mitigation policies for a given SSP. Mitigation costs do not only vary with the stringency of mitigation targets (RCP levels), but also with the degree of global cooperation on mitigation policy. See also the conceptualization of mitigation SPAs in a scenario matrix setting in Knopf et al. (2011; Figure 6).

The adaptation policy assumptions can also be varied across SPAs. Figure 2 shows the example of three different reduced SPAs with no, moderate and aggressive adaptation policies. In general, a SPA will include a consistent set of assumptions on mitigation and adaptation policies specifying, e.g., the degree of coordination of regional and sectoral mitigation efforts and the aggressiveness of adaptation measures. Combining the information in Figs. 1 and 2 will allow to compare total climate policy costs with residual climate damages for a given combination of SSP, RCP and SPA, and also will allow the exploration of interactions of adaptation and mitigation policies because the extent of efforts required to adapt later in the century will depend on mitigation policies implemented in the near term

As discussed in O'Neill et al. (2014), shared socioeconomic reference pathways should be chosen in a way to cover different levels

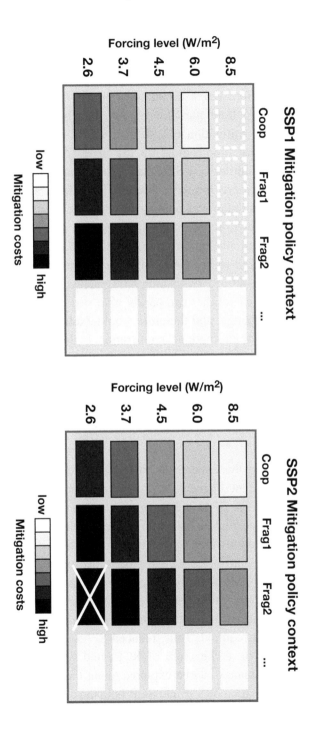

Figure 1. A policy axis can be added to the matrix architecture to explore how the costs of mitigation policy depend on assumptions regarding the form of mitigation action. Here, the costs assuming cooperative action (Coop) are compared to policies with different degrees of fragmented participation (Frag1 and Frag2) for SSP1 and SSP2 (see O'Neill et al. for a description of the SSP space). Some targets cannot be achieved with fragmented participation (indicated by the cross).

of socioeconomic challenges to mitigation and adaptation. Here we want to point out that SPAs will play an important role in translating those challenges into costs, benefits and obstacles to climate policy. For example, a SPA that foresees a group of countries that never adopt mitigation policies throughout the 21st century, can imply a greater difficulty to reach global mitigation targets than a SPA with global participation in a climate mitigation regime for the same SSP with given challenges to mitigation. Similarly, a SPA that restricts adaptation to domestic action will have a more limited scope for adaptation action than a SPA that allows for international pooling of adaptation resources. In many cases, the challenges to mitigation and adaptation in SSPs and the obstacles to mitigation and adaptation in SPAs will be correlated by the requirement that SPAs need to be consistent with the underlying SSPs (see above). For example, an SSP describing a fragmented world with regional blocks neutralizing each other can hardly give rise to global cooperative action on climate change, and therefore would exclude SPAs characterized by a large degree of international cooperation. In this situation, the challenges to mitigation and adaptation in the SSP would be augmented by the obstacles from the non-cooperative nature of the SPA.

11.6 SUMMARY AND CONCLUSIONS

We presented the concept of shared climate policy assumptions (SPAs) as an important element of the new scenario framework for climate change analysis. SPAs are the glue that allows the variety of alternative socioeconomic evolutionary paths to be coupled with the library of climate model simulations that were created using the RCPs. SPAs capture key climate policy attributes such as targets, instruments and obstacles. In their reduced form, SPAs are restricted to information that is neither specified in the socioeconomic reference pathways (SSPs), nor in the RCPs which largely determine the global outcome of mitigation action and the extent of adaptation required. Thus, reduced SPAs add a third dimension to the scenario matrix architecture of RCPs and SSPs. We demonstrated how the scenario matrix can be explored along the dimensions of SPAs. For a given combination of SSP and RCP, both the climate policy costs, including

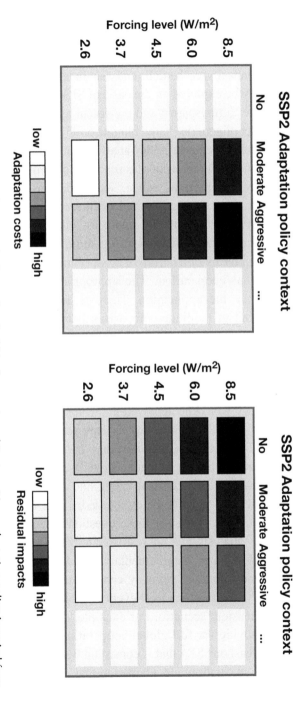

Figure 2. The type of adaptation policy can also be explored within the matrix architecture. Here, adaptation policy is varied from no adaption to moderate to aggressive adaptation. The matrix allows for the comparison of the costs and benefits of the adaptation policy for a given SSP, for example SSP2. The shading for climate policy adaptation costs and residual climate impacts is provided for illustrative purposes only.

adaptation and mitigation, as well as the residual climate damages will vary with the climate policy assumptions. Of course, elements other than costs and benefits are needed for a full policy appraisal (e.g. regional context, institutional capacity, co-benefits and risk trade-offs; McCollum et al. 2011), and those may be explored on the basis of SPAs as well. In summary, SPAs can contribute to transparent and consistent assumptions about policies. A meaningful set of generic shared climate policy assumptions will be needed to group individual climate policy analyses and facilitate their comparison.

SPAs should include assumptions about both mitigation and adaptation policies. Key assumptions relate, inter alia, to the degree of global cooperative action on mitigation and the international pooling of adaptation resources. SPAs will typically comprise quantitative information, e.g. relating to the regional distribution of mitigation effort, and qualitative information in terms of a narrative or storyline, e.g. relating to the degree of international cooperation. The appropriate level of detail of SPAs will depend on the application. There is an inherent tension between sufficient detail to facilitate comparison between climate policy studies and generality to group them into broad classes of SPAs.

SPAs should only contain information about climate policies, while other policies under consideration should become part of SSPs. The motivation of a policy may be used to determine whether it belongs in the SSP or the SPA. If a policy would be deployed in the absence of climate change, then it belongs in the SSP. If a policy is deployed and/or tightened solely in response to climate change, then it belongs in the SPA, but only that portion that is motivated solely by climate change. We acknowledge that for some multi-objective policies a clear dividing line does not always exist. Some decisions will have to be made on a case by case basis depending on the application context. However, these questions are part and parcel of any climate policy analysis, and not restricted to the use of SPAs.

While SPAs are complementary to SSPs, the character of a socioeconomic reference pathway can constrain the choice of plausible climate policy assumptions. Not every combination of SPA and SSP will provide a consistent scenario. For example, a fragmented world will not be able to provide for full global cooperation on climate change. There should be at least one SPA that is consistent with each SSP. Socioeconomic challenges to mitigation and adaptation that are described by an SSP can be augmented by the choice of SPA. For example, if global cooperation is highly constrained in a given SPA, this will add to the challenges to mitigation and adaptation in the SSP.

We conclude that SPAs are an important concept to facilitate climate policy analysis in the new scenario framework. Their formulation and application will have to be further developed and tested, which may involve iteration between modelers, analysts and various stakeholders. If SPAs are designed to be broad enough to allow an exploration of the relevant climate policy space, their use will likely provide new insights into the implications of alternative policy designs for climate action.

FOOTNOTES

1. We need to distinguish between the four RCPs created for climate model simulations (van Vuuren et al. 2011) and climate forcing used more broadly to characterize the rows of the scenario matrix (van Vuuren et al. 2014). Since the four RCPs establish the direct link to the climate model results, they are obviously prime candidates to define the rows of the matrix. However, additional rows for intermediate forcing levels such as 3.7 W/m2 could be added, or the forcing outcome could be described more broadly by the 2100 level of the RCPs rather than the entire pathway. A forcing pathway may deviate from an RCP in some aspect and still be associated with the matrix row defined by the RCP. Such forcing pathways have been called RCP replications.

2. On the other hand, it will sometimes prove useful to compare the effects of a consistent SPA with realistic policy representation with an "idealized policy" SPA that may not be fully consistent with the SSP in order to gain analytical insights.

REFERENCES

1. Babiker MH, Eckaus RS (2007) Unemployment effects of climate policy. Environ Sci Policy 10(7–8):600–609

2. Blanford G, Kriegler E, Tavoni M (2014) Harmonization vs. Fragmentation: overview of climate policy scenarios in EMF27. Clim Chang. doi:10.1007/s10584-013-0951-9

3. Calvin K, Clarke L, Krey V, Blanford G, Jiang K, Kainuma M, Kriegler E, Luderer G, Shukla PR (2012) The role of Asia in mitigating climate change: results from the Asia modeling exercise. Energy Econ 34(Supplement 3):S251–S260

4. Calvin K, Wise M, Kyle P, Patel P, Clarke L, Edmonds J (2014) Trade-offs of different land and bioenergy policies on the path to achieving climate targets. Clim Chang. doi:10.1007/s10584-013-0897-y

5. Clapp C, Ellis J, Benn J, Corfee-Morlot J (2012) Tracking climate finance: What and how? Paris, Organisation for Economic Co-operation and Development. http://www.oecd.org/env/climatechange/50293494.pdf

6. Clarke L, Edmonds J, Krey V, Richels R, Rose S, Tavoni M (2009) International climate policy architectures: overview of the EMF 22 International Scenarios. Energy Econ 31:S64–S81

7. Ebi et al (2014) A new scenario framework for climate change research: background, process, and future directions. Clim Chang. doi:10.1007/s10584-013-0912-3

8. Guivarch C, Crassous R, Sassi O, Hallegatte S (2011) The costs of climate policies in a second best world with labour market imperfections. Clim Pol 11:768–788

9. Hallegatte S, Przyluski V, Vogt-Schilb A (2011) Building world narratives for climate change impact, adaptation and vulnerability analyses. Nat Clim Chang 1(3):151–155. doi:10.1038/nclimate1135

10. Knopf B, Luderer G, Edenhofer O (2011) Exploring the feasibility of low stabilization targets. Wiley Interdisc Rev Clim Chang 2(4):617–626

11. Kriegler E, O'Neill BC, Hallegatte S, Kram T, Lempert R, Moss R, Wilbanks T (2012) The need for and use of socio-economic scenarios for climate change analysis: a new approach based on shared socio-economic pathways. Glob Environ Chang 22:807–822

12. Kriegler E et al (2014a) The role of technology for climate stabilization: overview of the EMF 27 study on global technology and climate policy strategies. Clim Chang. doi:10.1007/s10584-013-0953-7

13. Kriegler E et al (2014b) Making or breaking climate targets: the AMPERE study on staged accession scenarios for climate policy. Technol Forecast Soc Chang. doi:10.1016/j.techfore.2013.09.021

14. Linnér BO, Mickwitz P, Román M (2012) Reducing greenhouse gas emissions through development policies: a framework for analysing policy interventions. Clim Dev 4(3):175–186. doi:10.1080/17565529.2012.698587

15. Luderer G, Bertram C, Calvin K, De Cian E, Kriegler E (2014) Implications of weak near-term climate policies on long-term mitigation pathways. Clim Chang. doi:10.1007/s10584-013-0899-9

16. McCollum DL, Krey V, Riahi K (2011) An integrated approach to energy sustainability. Nat Clim Chang 1:428–429

17. Morgan MG, Keith DW (2008) Improving the way we think about projecting future energy use and emissions of carbon dioxide. Clim Chang 90:189–215

18. Moss RH et al (2008) Towards new scenarios for analysis of emissions, climate change, impacts, and response strategies. Intergovernmental Panel on Climate Change, Geneva, 132 pp
19. Moss RH et al (2010) The next generation of scenarios for climate change research and assessment. Nature 463:747–756
20. Nakicenovic N et al (2000) IPCC Special Report on Emissions Scenarios (SRES). Cambridge University Press, Cambridge
21. O'Neill BC et al (2014) A new scenario framework for climate change research: the concept of shared socio-economic reference pathways. Clim Chang. doi:10. 1007/s10584-013-0905-2
22. Parson EA (2008) Useful global change scenarios: current issues and challenges. Environ Res Lett 3, 045016
23. Riahi K et al (2014) Locked into copenhagen pledges—implications of short-term emission targets for the cost and feasibility of long-term climate goals. Technol Forecast Soc Chang. doi:10.1016/j.techfore.2013.09.016
24. Tavoni M et al (2014) The distribution of the major economies' effort in the Durban platform scenarios. Clim Chang Econ 4(4):1340009
25. van Vuuren DP et al (2011) The representative concentration pathways: an overview. Clim Chang 109:5–31
26. van Vuuren DP, Riahi K, Moss R, Edmonds J, Thomson A, Nakicenovic N, Kram T, Berkhout F, Swart R, Janetos A, Rose SK, Arnell N (2012) A proposal for a new scenario framework to support research and assessment in different climate research communities. Glob Environ Chang 22:21–35
27. van Vuuren DP et al (2014) A new scenario framework for climate change research: scenario matrix architecture. Clim Chang. doi:10.1007/s10584-013-0906-1
28. Winkler H, Höhne N, Den Elzen M (2008) Methods for quantifying the benefits of sustainable development policies and measures (SD-PAMs). Clim Pol 8(2):119–134
29. Winkler H, Vorster S, Marquard A (2009) Who picks up the remainder? Mitigation in developed and developing countries. Clim Pol 9(6):634–651

AUTHOR NOTES

CHAPTER 1

Acknowledgments
I would like to thank colleagues at Landcare Research, especially Phil Cowan, for discussions underlying the development of the proposed methodology, and Robbie Andrew, Anne Austin, Annette Cowie, Édouard Périé and Katsumasa Tanaka and anonymous reviewers for many useful and insightful comments on the manuscript.

CHAPTER 2

Author Contributions
All the authors designed the research and performed some the research and preliminary analyzed; Lucian-Ionel Cioca and Larisa Ivascu wrote the paper. All authors contributed to a deeper data analysis, read and approved the final manuscript.

Conflicts of Interest
The authors declare no conflict of interest.

CHAPTER 4

Acknowledgments
Air temperature data for Athens were provided by the Institute for Environmental Research and Sustainable Development of the National Observatory of Athens. This work has been supported by grant no by the Norwegian Research Council.

Author Contributions

Isaksen has carried out the main part of writing, although all coauthors have contributed to the text. The research has mainly been carried out by the co-authors, comprising both modelling work and also data analysis.

Conflicts of Interest

The authors declare no conflict of interest.

CHAPTER 5

Acknowledgments

The study was supported by the National Science Foundation (ATM07-21142) and University of California, San Diego Open Access Fund (Pilot). We thank Nathan BorgfordParnell, Stephen O. Andersen, and Dennis Clare for reading the manuscript; and Allison Thomson and Keywan Riahi for sharing data. We acknowledge two anonymous reviewers and Steven J. Smith for their constructive comments that greatly improved the paper.

CHAPTER 6

Acknowledgments

Yuefen Gao, Honglei Zhao and Yingxin Peng are grateful to the financial support by the Fundamental Research Funds for the Central Universities (Project No: 11MG43). This work is also partly supported by European Regional Development Fund (ERDF 2009-12).

CHAPTER 7

Acknowledgments

EAD acknowledges support from the NSF Research Coordination Network award DEB-1049744. The authors thank E Suddick and G Fisk for assistance with figure 2 and S Leonard for assistance with figure 3. We also thank other team members of the UNEP report.

CHAPTER 8

Acknowledgments

The authors wish to thank the coordinators and students of the Industrial Ecology education program at Leiden University and Delft University of Technology for their helpful and incisive comments on a presentation of an early version of this paper. The authors would also like to thank two anonymous reviewers for their valuable comments on a previous draft of the paper.

CHAPTER 9

Conflicts of Interest

The authors declare that there is no conflict of interests regarding the publication of this paper.

Acknowledgments

This work is supported by National Nature Science Foundation of China under Grants 61372195, 61371200, and 61304069.

CHAPTER 10

Acknowledgments

This paper has been written as part of the COMBINE (grant agreement no.: 226520) and PEGASOS (grant agreement no: 265148) projects, both funded by the European Commission within the Seventh Framework Programme.

CHAPTER 11

Acknowledgments

Jae Edmonds' participation was supported by the Integrated Assessment Research Program in the Office of Science, U.S. Department of Energy.

INDEX